Pentoses and Lignin

Th. W. Jeffries

Utilization of Xylose by Bacteria, Yeasts, and Fungi

L. D. McCracken, Ch.-Sh. Gong

D-Xylose Metabolism by Mutant Strains of *Candida* sp.

H. Schneider, R. Maleszka, L. Neirinck, I. A. Veliky, P. Y. Wang, Y. K. Chan

Ethanol Production from D-Xylose and Several Other Carbohydrates by *Pachysolen tannophilus* and Other Yeasts

C. P. Kurtzman

Biology and Physiology of the D-Xylose Fermenting Yeast *Pachysolen tannophilus*

N. B. Jansen, G. T. Tsao

Bioconversion of Pentoses to 2,3-Butanediol by *Klebsiella pneumoniae*

B. Volesky, T. Szczesny

Bacterial Conversion of Pentose Sugars to Acetone and Butanol

H. Janshekar, A. Fiechter

Lignin — Biosynthesis, Application, and Biodegradation

AKADEMIE-VERLAG · BERLIN

Pentoses and Lignin

Pentoses and Lignin

Managing Editor: A. Fiechter

with 46 Figures and 46 Tables

Akademie-Verlag · Berlin 1983

Die Originalausgabe erscheint im Springer-Verlag
Berlin—Heidelberg—New York
als Volume 27 der Schriftenreihe Advances in Biochemical Engineering/Biotechnology

Vertrieb ausschließlich für die DDR und die sozialistischen Länder

Erschienen im Akademie-Verlag, DDR-1086 Berlin, Leipziger Straße 3—4
Lizenznummer: 202 · 100/540/83
Printed in the German Democratic Republic
Gesamtherstellung: VEB Druckerei „Thomas Müntzer“, 5820 Bad Langensalza
Umschlaggestaltung: Karl Salzbrunn
LSV 1315, 1345
Bestellnummer: 763 275 0 (6789)
08600

Table of Contents

Utilization of Xylose by Bacteria, Yeasts, and Fungi

Thomas W. Jeffries*
Microbiologist. Forest Products Laboratory, U.S. Dept. of Agriculture,
P.O. Box 5130, Madison, Wisconsin 53705, U.S.A.

Hemicellulosic sugars, especially D-xylose, are relatively abundant in agricultural and forestry residues. Moreover, they can be recovered from the hemicelluloses by acid hydrolysis more readily and in better yields than can D-glucose from cellulose. These factors favor hemicellulosic sugars as a feedstock for production of ethanol and other chemicals. Unfortunately, D-xylose is not so readily utilized as D-glucose for the production of chemicals by microorganisms. The reason may lie in the biochemical pathways used for pentose and hexose metabolism. Different pathways are employed by prokaryotes and eukaryotes in the initial stages of pentose assimilation. Transport and phosphorylation possibly limit the overall rate of D-xylose utilization. The intermediary steps of pentose metabolism are generally similar for both bacteria and fungi, but substantial variations exist. Phosphoketolase is present in some yeasts and bacteria able to use pentoses. Regulation of the oxidative pentose phosphate pathway occurs at D-glucose-6-phosphate dehydrogenase by the intracellular concentration of NADPH. Regulation of nonoxidative pentose metabolism is not well understood. In some

* Maintained in cooperation with the University of Wisconsin.

yeasts and fungi, conversion of D-xylose to ethanol takes place under aerobic or anaerobic conditions with rates and yields generally higher in the former than in the latter. Xylitol and acetic acid are major byproducts of such conversions. Many yeasts are capable of utilizing D-xylose for the production of ethanol. Direct conversion of D-xylose to ethanol is compared with two-stage processes employing yeasts and D-xylose isomerase.

1 Introduction

Hemicellulosic sugars in acid hydrolysates of hardwoods and agricultural residues could become important feedstocks for the production of ethanol and other chemicals by microbial processes. Several factors favor their use: they are relatively abundant in a variety of common lignocellulosic residues; they can be recovered by mild acid hydrolysis; and new microbiological processes are being developed for their conversion.

Within the past 2 years, several significant findings have advanced the prospects for production of ethanol and other chemicals from D-xylose. First, yeasts, which were previously considered unable to ferment[1] 5-carbon sugars, have now been shown

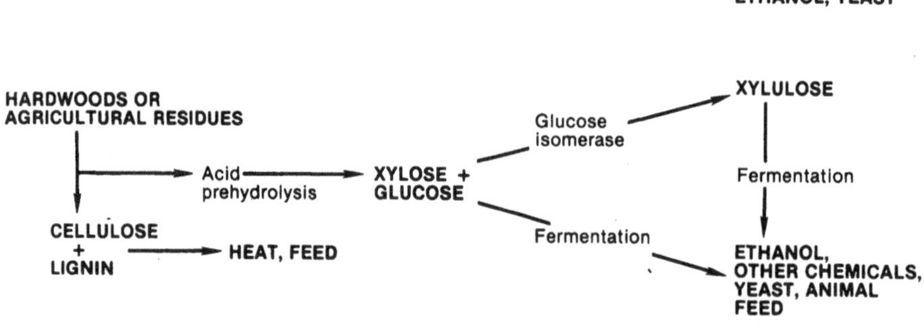

Advances i. Bio. Eng.

Fig. 1. Comparison of ethanol production from grain and lignocellulosic residues. (M 151671). For definition see footnote

[1] The term "fermentation" and its various derivatives is used herein to refer to dissimilatory metabolic processes through which an organic substrate is converted into oxidized and reduced products without a net overall change in the oxidation state.

to utilize the pentulose D-xylulose under anaerobic conditions [1-5]. Since D-xylulose can be formed from D-xylose through the action of glucose isomerase (actually xylose isomerase) [6,7], processes have been developed employing two-stage isomerization and fermentation [8-12]. Second, several yeasts, particularly those belonging to the genera *Pachysolen* [13-15] and *Candida* [16-18], have been shown to convert D-xylose to ethanol under aerobic and anaerobic conditions. Besides these recent findings, it is known that certain fungi, particularly *Fusarium lini* [19-23], are capable of converting D-xylose to ethanol; and various bacteria can form several potentially useful products such as ethanol, acetic acid, 2,3-butanediol, acetone, isopropanol, and n-butanol from D-xylose [24-29]. Finally, unlike the fermentation of sugar from grains, utilization of pentoses derived from forestry and agricultural residues for the production of chemicals does not decrease food supplies; indeed, by virtue of the production of microbial biomass and unutilized sugars, such processes can supplement animal feed resources (Fig. 1).

This review attempts to examine recent microbiological findings in relation to the previous understanding of D-glucose fermentation and pentose metabolism. It briefly examines the availability of hemicellulosic sugars — particularly D-xylose — in lignocellulosic residues, reviews aspects of pentose metabolism and metabolic regulation of fermentative processes, and discusses some recent research progress on aerobic and anaerobic conversions of D-xylose to ethanol by yeasts and bacteria. The 2,3-butanediol fermentation and the butanol/acetone/ethanol fermentations are reviewed in other chapters of this volume.

1.1 Distribution of Pentoses in Lignocellulosic Residues

Hemicelluloses are widely distributed, major components of lignocellulosic materials comprised of neutral sugars, uronic acids, and acetal groups, all present as their respective anhydrides (e.g., the anhydride of D-xylose is xylan). As the anhydrides, hemicellulosic sugars average about 26% of the dry weight of hardwoods,[2] and 22% of softwoods and about 25% of several major agricultural residues. Pectin, ash, and protein account for variable fractions in lignocellulosic materials, whereas cellulose (anhydro D-glucose) and lignin make up the balance (Table 1).

The xylan and arabinan contents of hemicelluloses vary with the plant species. The xylan content of hardwoods is generally much higher than that of softwoods, ranging between 11% and 25% in the former and between 3% and 8% in the latter [30-32]. Hemicelluloses in hardwoods contain appreciable amounts of D-xylose, D-mannose, acetyl, and uronic acid. The acetyl content ranges between 3% and 4.5% in hardwoods and between 1% and 1.5% in softwoods; uronic acid (as the anhydride) ranges between 3% and 5% in both hardwoods and softwoods [30]. In conifers, the predominant hemicellulosic sugar is D-mannose, which, as mannan, averages about 11% of the total dry weight [32]. Whereas the xylan content of softwoods is lower than in hardwoods, the lignin content is higher. The predominant hemicellulosic sugar of agricultural residues is D-xylose. The xylan content of corn residues varies

[2] The word "hardwoods" refers to broad-leafed trees (angiosperms) and has nothing to do with the hardness of the woods. Similarly, "softwoods" refers to coniferous trees (gymnosperms).

from about 17% in the leaves and stalks to 31% in the cobs, but, on the average, it comprises about 24% of the total dry weight of corn stover (L. H. Krull, personal communication). The chemistry of the hemicelluloses of grasses has been reviewed recently [35].

Table 1. Proximate composition of various biomass resources

	% of total dry weight									
	Glucan	Galactan	Mannan	Arabinan	Xylan	Hemicellulosic sugars[a]	Hemicellulose[b]	Cellulose[c]	Lignin[d]	Ref.
Hardwoods	50	0.8	2.5	0.5	17.4	26.2	34	45	21	30–32)
Softwoods	46	1.4	11.2	1.0	5.7	22.3	28	43	29	32–35)
Wheat straw	35	0.7	0.4	4.4	19	28.5	—	31	14	33–34)
Corn stalks	36.5	1.1	0.6	2.1	17.2	27.5	—	30	—	33–34)
Soybean residue	38	1.8	2.4	1.0	12.5	18.7	—	37	—	33–34)

[a] Reported as anhydrides;
[b] Includes acetyl- and uronic-acid residues;
[c] Residual glucan following acid prehydrolysis;
[d] Analyzed as Klason lignin (acid-insoluble)

Overall, it would appear that the high hexose (D-glucose plus D-mannose) content of softwoods would favor their utilization as fermentation feedstocks. Presently, however, most softwood residues find their way into pulping operations because of the favorable fiber characteristics of conifer species. In contrast, hardwood residues have much less value for paper production and are generally burned for the generation

Table 2. Estimated and projected total agricultural residues in the United States [36,37]

Material	Quantity	
	1980	2000
	10⁶ ODT	
Corn residue	100	142
Wheat straw	101	87
Soybean residue	98	159
Other grains	57	74
Other agricultural products	28	35
Totals	385	497

of process heat. Most wood residues and agricultural residues are not generally collected at the time of harvest. Branches, leaves, tops, and roots of trees are left in the forest when the merchantable bole is taken for pulp or lumber production; only 50 to 75% of the tree is removed during harvest [39–40]. More residues are generated during milling and pulping operations. A certain proportion of agricultural residues (about 1 ton per acre) are left in the soil to maintain tilth and prevent erosion [33]. A notable exception is sugarcane bagasse. In this instance, the residues are recovered at the sugar mill where they are burned to provide heat for sugar processing. Such collection greatly facilitates economic utilization. Wood mill residues hold a similar advantage.

Of all biomass resources in the United States, low-grade hardwoods and agricultural residues are the two largest available components (Tables 2 and 3). Each possesses

Table 3. United States forest biomass

Category	Quantity	
	Total (Ref.)	Available (Ref.)
	10^6 ODT a^{-1}	
Harvest sites:		
Above ground	160 [38]	110 [41]
Stumps and roots	50 [39]	
Residues:		
Mills (wood)	65 [39]	12 [39,41]
(bark)	18 [39]	8 [39]
Urban tree removals	70 [38]	35[a]
Land clearing	20 [38]	10[a]
Commercial forest lands:		
Surplus growth	270 [40]	150[a]
Annual mortality	95 [38]	50[a]
Noncommercial forest lands:		
Reserved	25 [41]	0
Unproductive	27 [41]	0
Totals	800	375

[a] Author's estimate

characteristics that favor its utilization. Low-grade hardwoods are abundant in the southeastern United States and, aside from direct combustion, have few commercial uses. They can be harvested on a year-round basis using presently available technology. Some hemicellulosic sugars are available as a byproduct of hardboard and insulation board manufacture; others are available as a byproduct formed during the manufacture of sulfite and dissolving pulps.

Significant quantities of wood residues are combusted for the production of steam [42]; wood and bagasse supplied $4 \sim 6 \times 10^{14}$ kJ to the United States' energy budget in 1980. Even though the hemicellulose comprises up to 30% by weight of these materials, it has only about two-thirds of the heat value of the lignin. Hence it is possible to remove the hemicellulosic sugars for use as a fermentation feedstock and combust the cellulose and lignin residues with relatively little loss of the energy value from the original feedstock.

1.2 Recovery of Hemicellulosic Sugars

Unlike cellulose, which is impermeable even to water, hemicellulose has a relatively open structure. This molecular architecture facilitates diffusion of acid into the polymer and speeds hydrolysis. Moreover, hemicellulose facilitates its own hydrolysis: Acetyl groups are readily hydrolyzed off, and the resulting acetic acid catalyzes the partial depolymerization of the hemicellulose.

In general, hemicellulosic sugars can be recovered with milder treatment and in better yield than can glucose from cellulose[43)]. Research at the Forest Products Laboratory in Madison, Wisconsin, has shown that more than 80% of the D-xylose can be recovered from southern red oak (*Quercus falcata*) wood chips through dilute sulfuric acid hydrolysis; in contrast, it is unlikely that more than 50% of the total D-glucose can be recovered from the residual cellulose in a second-stage acid hydrolysis, which must be carried out at higher temperature (Fig. 2). The significance of this finding is that about half of the sugars produced from wood by a two-stage acid hydrolysis process are from the hemicellulose.

In the case of corn residues the situation is even more extreme. Cellulose and hemicellulose each make up approximately 30% of the total dry weight of the material. Through dilute (0.8%–1.2%) sulfuric acid hydrolysis carried out at relatively

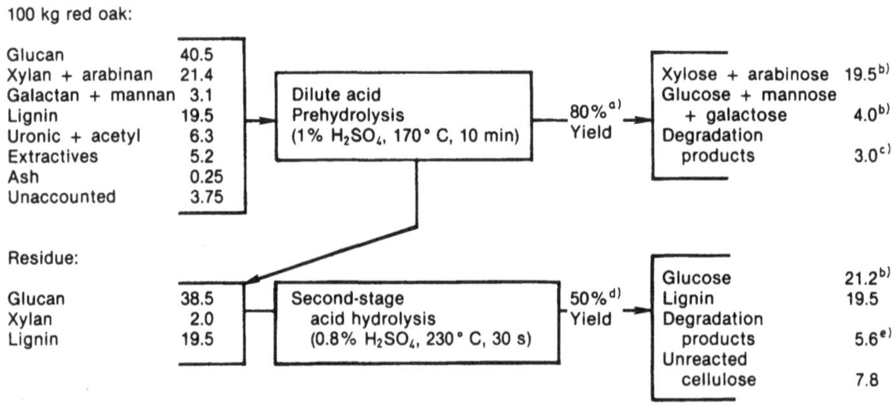

Fig. 2. Dilute acid hydrolysis of southern red oak (*Quercus falcata*). Data are taken from pilot- and bench-scale experiments at the Forest Products Laboratory, Madison, Wis.

[a] Prehydrolysis yield is based on the amount of D-xylose recovered versus the potential amount present as the anhydride polymer, xylan, in the original feedstock.

[b] Products are expressed as free sugars.

[c] Degradation products are mainly furfural.

[d] Second-stage acid hydrolysis yield is based on the amount of D-glucose recovered versus the potential amount in the prehydrolysis residue.

[e] Degradation products are mainly hydroxymethyl furfural. (M 151672)

mild temperature (100 C), more than 90% of the hemicellulosic sugars can be recovered by pressing and washing after 30 min [43]. Yields of D-glucose from the residual cellulose in a second, necessarily more drastic, hydrolysis are appreciably lower.

Two-stage dilute acid hydrolysis has been developed to enhance the recovery of pentose sugars [44], and the partial acid hydrolysis that releases the pentoses has been found to be favorable as a pretreatment for subsequent enzymatic saccharification of the residual cellulose [45]. One of the principal disadvantages of acid hydrolysis is that, even under relatively mild conditions, appreciable amounts of furfural and hydroxymethylfurfural are formed from D-xylose and D-glucose. These dehydration products are generally toxic to yeasts [46]. However, the concentration of furfural decreases during fermentation of glucose by *Saccharomyces* [47], indicating that detoxification by the organism is possible.

While it is conceivable that economical enzymatic or chemical processes might be developed to hydrolyze the cellulosic residue in the second stage, utilization of this residue for sugar production is not always essential for the economic practicability of a first-stage hydrolysis. As mentioned above, dilute sulfuric acid hydrolysis to remove hemicellulosic sugars would still leave a significant fuel component. It is also possible that acid-prehydrolyzed wood might be incorporated into cellulosic products.

For the purposes of this review, it is sufficient to note that pentoses are presently available as byproducts of industrial processes such as fiberboard manufacture and sulfite pulping, and that supplies of these sugars could be significantly increased. Conversion of pentoses to ethanol and other chemicals can be accomplished through microbial processes. To better understand the rate-limiting step(s) in this conversion, the biochemistry of D-xylose utilization is treated in the following section.

2 D-Xylose Metabolism

The biochemical mechanisms of D-xylose metabolism are quite different from those for D-glucose. Whereas D-glucose is metabolized by the Embden-Meyerhoff-Parnas pathway, D-xylose metabolism proceeds by way of the pentose phosphate pathway (PPP). Following transport into the cell, D-xylose is either isomerized or reduced, then reoxidized to form D-xylulose. This sugar is then phosphorylated, isomerized, and rearranged to form a metabolic pool of phosphorylated 3-, 4-, 5-, 6-, and 7-carbon sugars at equilibrium within the cell. The PPP interacts with the Embden-Meyerhoff-Parnas pathway and other parts of intermediary metabolism. D-Glucose can enter the PPP through either oxidative or nonoxidative reactions. Intermediates can exit the PPP through the formation of nucleic acids, aromatic amino acids, lipids, and other metabolic end products.

The objective of this section is to examine the principal steps of D-xylose metabolism in prokaryotes and eukaryotes with the aim of elucidating those reactions that might limit the overall rate of D-xylose utilization. More general aspects of pentose and pentitol metabolism have been reviewed elsewhere [48,49].

2.1 Transport

Transport across the cell membrane is the first step in the metabolism of D-xylose or any other nutrient; as in the cases of D-glucose, it can limit the overall rate of utilization [50]. Sugar transport can occur by at least three mechanisms: passive (or physical) diffusion, facilitated diffusion, or active transport. Active transport can be further classified into chemiosmotic, direct energization, and group translocation mechanisms [51]. These processes are not mutually exclusive and two or more may function in a single organism.

Passive diffusion is the simplest and slowest process. It requires a substantial concentration gradient and progresses by diffusion of the solute across the plasma membrane. Only small, lipid-soluble molecules such as glycerol or ethanol are transported by this mechanism to any appreciable extent. The rate of transport for most sugars by passive diffusion is probably negligible, but the transport of acyclic polyols (erythritol, xylitol, ribitol, D-arabinitol, mannitol, sorbitol, and galactitol) has been reported to occur by such a mechanism in *Saccharomyces cerevisiae*. As is characteristic of passive diffusion, the process is independent of pH, uncoupling agents, and uranyl ions, and the initial rate of transport increases with the solute concentration [52].

Facilitated diffusion, like passive diffusion, requires no metabolic energy, employs a concentration gradient, and attains an equilibrium of sugar concentrations inside and outside the cell. Unlike passive diffusion, however, it is mediated by a carrier protein and hence exhibits specificity toward the substrate molecule. Although the rates of both passive and facilitated diffusion increase with increasing sugar concentrations, facilitated diffusion is distinguished by the fact that the system becomes saturated at some concentration, resulting in a maximum rate of transport. Other characteristics of facilitated diffusion are that similar sugars competitively inhibit transport, that uranyl ions specifically inhibit transport and that facilitated diffusion exhibits counter transport. Facilitated diffusion is used by many types of eukaryotic cells for transport of sugars.

Active transport mechanisms, like facilitated diffusion, are mediated by carrier proteins, and hence exhibit the properties of saturability, substrate specificity, and specific inhibition, but the processes require metabolic energy and can transport sugars against a concentration gradient. Metabolic energy can be provided by establishing a membrane potential as in the chemiosmotic mechanism, by the hydrolysis of adenosine 5′-triphosphate (ATP) as in the direct energization mechanism, or by the transfer of phosphate from phospho*enol*pyruvate (PEP) to the sugar substrate as in the group translocation mechanism.

2.1.1 Bacteria

Bacteria generally employ active transport mechanisms for the uptake of sugars and other nutrients. Transport of D-xylose across the cell membrane of *Escherichia coli* is linked to the movement of protons, as evidenced by a rise in pH in the extracellular medium upon addition of D-xylose to an energy-depleted suspension of cells [53]. This evidence supports the chemiosmotic symport mechanism proposed by Mitchell [54] in which protons and D-xylose are transported together across the cell membrane. In *E. coli*, accumulation of [^{14}C] D-xylose is inhibited by various uncoupling agents such as tetrachlorosalicylamide (TCS), 2,4-dinitrophenol (DNP) and carbonylcyanide *m*-chlorophenylhydrazone (CCCP), which destroy the proton gradient. Neither sodium fluoride, which prevents PEP formation by the enolase reaction, nor arsenate, which drastically reduces the intracellular concentration of ATP, inhibits D-xylose transport, implying that it is energized by a chemiosmotic mechanism and not by directly energized or PEP-phosphotransferase mechanisms [53]. The PEP-phosphotransferase system is not involved in the uptake of D-xylose and xylitol by *Staphylococcus xylosus* and *Staphylococcus saprophyticus* either [55], but pentitols (ribitol and xylitol) are transported by a substrate-specific PEP-phosphotransferase system in *Lactobacillus casei* [56,57]. In contrast to D-xylose transport in which an energized membrane state is employed, the D-ribose transport system of *E. coli* is apparently coupled to the hydrolysis of ATP [58,59].

The D-xylose transport of *E. coli* is relatively specific as indicated by the fact that L-arabinose, D-ribose, D-lyxose, xylitol, and D-fucose fail to promote pH changes in D-xylose-induced cells [53]. The specificity of D-xylose transport is somewhat lower in *Salmonella typhimurium*. In this organism, L-arabinose accumulates in D-xylose-induced cells and D-xylose accumulates in L-arabinose-induced cells. Xylitol and L-arabinose compete against D-xylose uptake, but D-arabinose, D-lyxose, and L-lyxose do not [60].

2.1.2 Yeasts and Fungi

In yeasts, D-xylose transport can occur by either facilitated diffusion or active processes. In *Saccharomyces cerevisiae*, D-xylose is nonmetabolizable and it appears to be transported by a facilitated diffusion process [61]. Transport of D-xylose is apparently related to D-glucose transport because, in the presence of D-glucose, the influx of D-xylose or D-arabinose is more rapid under anaerobic than under aerobic conditions. In the absence of D-glucose, the rate of entry of these two sugars is identical under either condition [61].

Transport of D-xylose by eukaryotes has been most extensively studied in *Rhodotorula* where it occurs by an active process [62–70]. Under aerobic conditions, D-xylose is accumulated against a concentration gradient of 1,000 with an apparent K_m of 2 mM. Under anaerobic conditions, transport is blocked, indicating that respiration is essential for transport in these obligately aerobic yeasts [61]. Alcorn and Griffin [62] have reported the presence of at least two carriers for D-xylose in the membrane of *Rhodotorula gracilis* (= *R. glutinis*). The carrier exhibiting a low K_m (high affinity) is repressed in rapidly growing cells and derepressed by starvation. Several hexoses competitively inhibit D-xylose transport, but the low K_m carrier exhibits greater specificity [62]. There is a single common system for D-xylose and D-galactose, but another distinct system for D-fructose. The transport of D-glucose has a special position in that D-glucose blocks all other systems observed, although D-glucose itself is transported by one of the systems [64]. D-xylose competitively inhibits the transport of D-glucosamine, indicating a single system for these two sugars [69]. Transport of nonmetabolizable monosaccharides by *R. gracilis* is partially inhibited by raising the temperature and greatly inhibited by uncouplers of oxidative phosphorylation [63]. Rotenone, antimycin A, potassium cyanide, sodium azide, oligomycin, dicyclohexylcarbodiimide (DCCD), DNP and CCCP all inhibit active transport in yeasts [68].

The uptake of monosaccharides and polyols by *R. gracilis* is accompanied by proton transport. Addition of D-xylose, D-glucose, 3-0-methyl-D-glucose or D-galactose (but not melibiose) to aqueous suspensions of *R. gracilis* causes a rapid increase in the extracellular pH. Similar but slower responses can be demonstrated with the addition of xylitol or ribitol to induced, but not uninduced, cells. The membrane potential, essential for proton mediated transport, is demonstrable by the intracellular accumulation of liquid-soluble cations and is strongly pH-dependent [67]. Höfer and Misra [65] have demonstrated that the intracellular steady-state concentration of D-xylose changes with the pH of the external medium and that the alteration is reversible. This reversibility can be explained by assuming that the carrier affinity is reversibly changed with external pH. At pH 8.5, the gradient vanishes and no D-xylose transport is demonstrable. The stoichiometry of H^+ and D-xylose uptake, determined under various physiological conditions, is one H^+ per sugar molecule taken up. The half-saturation constants of H^+ and sugar uptake are also similar. These data strongly support the hypothesis that D-xylose transport in *R. gracilis* is energized by an electrochemical gradient of H^+ across the plasma membrane and functions by the H^+ symport mechanism [65].

Active transport, at least for hexoses, also seems to be present in *Candida parapsilosis* [71] and *Candida guilliermondii* [72]. Uptake of D-xylose has also been studied in *Penicillium* and *Fusarium* species [73]. Other aspects of sugar transport in yeasts have been recently reviewed [51].

2.2 Conversion of D-Xylose to D-Xylulose-5-Phosphate

Once inside the cell, D-xylose is first converted to D-xylulose and then phosphorylated. A basic difference seems to exist between prokaryotes and eukaryotes in the initial metabolism: Bacteria generally employ an isomerase to convert D-xylose to D-xylulose [6,7,25], whereas yeasts and fungi carry out the same conversion through a two-step reduction and oxidation [74] (Fig. 3). A few exceptions to this generalization have, however, been reported [75,76]. Conversion of D-xylose to D-xylulose is apparently a critical step in yeasts. It has long been recognized that roughly half of all yeasts are capable of assimilating D-xylose under aerobic conditions [77], but, until recently, none were known to utilize D-xylose anaerobically. In contrast, the ketoisomer, D-xylulose, is used anaerobically by many yeasts [1-4,10], despite the fact that it is not found extracellularly in nature. The relative ease with which D-xylulose

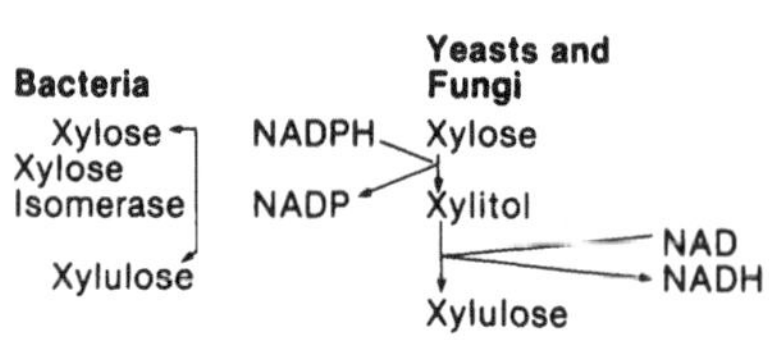

Fig. 3. Yeasts and bacteria generally employ different pathways for D-xylose assimilation

is utilized by yeasts suggests that the reductive/oxidative conversion of D-xylose to D-xylulose is regulated or rate-limiting in these organisms. The fact that many yeasts utilize D-xylose under aerobic conditions further suggests that this regulation is related to aerobic metabolism. Numerous bacteria employing D-xylose isomerase exhibit diverse patterns of D-xylose utilization under both aerobic and anaerobic conditions and yield a variety of end products [11,25-29]. The utilization of D-xylulose by bacteria has not been studied except in the case of *Zymomonas mobilis*, which did not use D-xylulose [10].

2.2.1 Isomerization

D-xylose isomerase, which catalyzes the reversible isomerization of D-xylose to D-xylulose or of D-glucose to D-fructose, has been well studied [78-84] and recently reviewed [6,7]. Comments herein will be limited to physiological aspects of importance to pentose utilization and to recent applications in the fermentation of D-xylose (via D-xylulose) by yeasts.

In the presence of D-xylose isomerase, at equilibrium, about 16% of the sugar is isomerized to D-xylulose [79]. This value increases with increasing temperature [12] and the presence of borate [12,79]. Maximum conversion (80%) of 1.0 M D-xylose to D-xylulose is observed in the presence of 0.2 M sodium tetraborate. Temperature and pH have no significant effect on the equilibrium value in the presence of this compound [12]. Temperature and pH optima for D-xylose isomerase depend on its source, and range between 50 and 90 C and pH 6.0 and 9.5, respectively [7]. Some D-xylose isomerases require cobalt for activity; almost all require Mg^{++}. Stability and half-life in commercial preparations depend greatly on the manner of immobilization [6,7,82]. Both D-xylose isomerase and D-xylulokinase are specifically induced by D-xylose in *Klebsiella pneumoniae* (= *Aerobacter aerogenes*) [85] and in other bacteria [7,55,60]. In *Staphylococcus*, utilization of D-xylose is inhibited by the presence of xylitol, due to the inhibition of D-xylose isomerase [55].

In *Klebsiella pneumoniae*, aldoses are isomerized and pentitols are dehydrogenated to form the corresponding pentuloses (Fig. 4). D-xylulose is the product of D-xylose, D-arabitol, D-lyxose, and xylitol. Other pentoses and pentitols form D- or L-ribulose or L-xylulose. These sugars are phosphorylated by kinases and the products D-ribose-5-phosphate, D- or L-ribulose-5-phosphate and L-xylulose-5-phosphate converge on D-xylulose-5-phosphate via isomerase and epimerases [48].

The presence of D-xylose isomerase has been reported in *Candida utilis* [75] and *Rhodotorula gracilis* [76]. Horitsu, Sasaki, and Tomoyeda [86] have also identified and partially purified an L-arabinose isomerase from *C. utilis*. The D-xylose isomerase of *C. utilis* was produced adaptatively when the organism was grown in a medium containing D-xylose. After partial purification, the enzyme was found to have a pH optimum of 6.5, a temperature optimum of 70 C, and a requirement for divalent cations, particularly Mn^{++}, Co^{++}, and Mg^{++} [75]. The D-xylose isomerase could be separated from the L-arabinose isomerase of *C. utilis* by ion-exchange chromatography on DEAE Sephadex A-50. The L-arabinose isomerase showed a pH optimum of 7.0, a temperature optimum of 60 C and stimulation of activity by Mn^{++} [86]. In the case of *R. gracilis*, Höfer, Betz, and Kotyk [76] concluded that induction of D-xylose isomerase was necessary for D-xylose metabolism. The assay for enzyme

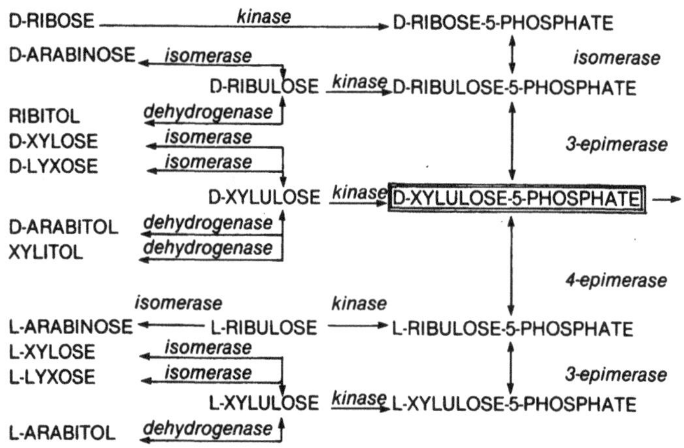

Fig. 4. Pathway of pentose and pentitol conversion to D-xylulose-5-phosphate in *Klebsiella pneumoniae* [48]. (M 151527)

activity was based on the disappearance of D-xylose from cell-free extracts of induced cells. No NADH or NADPH was added to the reaction mixture, and xylitol was not utilized as a carbon source even though it was taken up by the cells. These findings supported their conclusion that D-xylose isomerase and not D-xylose reductase was necessary for D-xylose assimilation in *R. gracilis*. All other investigations have indicated that these and other yeasts and filamentous fungi employ the reductive/oxidative route for D-xylulose formation [74]; consequently, these reports of D-xylose isomerase in yeasts require further confirmation.

2.2.2 Reduction, Oxidation, and Polyol Formation

The reduction of D-xylose to xylitol is catalyzed by alditol: NADP 1-oxidoreductase, also known as aldose reductase [87–92]. The purified enzyme is active with a wide variety of sugars but is most active on those in which the hydroxyl group at carbon-2 is in the D-glycero configuration [89]. An enzyme with preference for the alternative configuration has also been described [93]. During purification, multiple aldose reductases are found [92], sometimes with various different substrate and coenzyme specificities [94]. In virtually every instance, however, the enzyme catalyzing reduction of D-xylose to xylitol in yeasts and fungi has a specificity for NADPH [74,88–96]. The reoxidation of xylitol to D-xylulose is catalyzed by xylitol: NAD 2-oxidoreductase [74,88,91,97,98]. The reaction is readily reversible, and NADH oxidation occurs with a number of ketoses including D-xylulose, D-fructose, and D-ribulose. The purified enzyme shows no activity with NADP(H) [97].

Evidence for the existence of the reductive/oxidative pathway in fungi and yeasts has been presented for four species of *Aspergillus*, and three species of *Penicillium*, and for *Rhizopus nigricans*, *Gliocladium roseum*, *Byssochlamys fulva*, *Myrothecium verrucaria*, *Neurospora crassa* [74], *Sclerotinia sclerotiorum* [91], *Oospora lactis* [88], *Cephalosporium chrysogenus* [99], *Candida albicans* [100], *Candida pulcher-*

rima[101], *Candida tropicalis*[102], *Candida utilis*[87,90,94], *Rhodotorula*[74,103], *Torulopsis candida*[93,94], *Torulopsis utilis*[74], and *Pichia quercuum*[92]. In the case of *Pichia quercuum*, Suzuki and Onishi[104] have described the simultaneous production of D-xylonic acid and xylitol from D-xylose by cell-free extracts. D-xylose dehydrogenase activity was dependent on NADP and D-xylose reductase activity was dependent on NADPH. Together these two enzymes coupled through regeneration of NAD and NADPH.

Bacteria have been shown to possess both alditol: NADP oxidoreductase for the production of xylitol[105] and NAD-dependent pentitol dehydrogenases[106-109] for the assimilation of pentitols. Both NADPH and NADH-aldopentose reductase activities and NAD-pentitol dehydrogenase activities have been described in *Mycobacterium phlei* grown on glycerol[110], but it is not known whether this organism uses an isomerase or reductive/oxidative pathway for the utilization of D-xylose.

Bacteria have also been shown to employ some unusual pathways for the metabolism of D-xylose and xylitol[56,57,111,112]. An NAD-D-xylose dehydrogenase has been described in *Arthrobacter* sp.[111]. This enzyme catalyzes the following reaction:

$$\text{D-xylose} + \text{NAD}^+ \rightarrow \text{D-xylonolactone} + \text{NADH} + \text{H}^+ \qquad (1)$$

This enzyme has been purified from D-xylose-grown cells and shown to be specific for NAD and D-xylose[112]. The lactone is subsequently hydrolyzed to xylonic acid.

Certain strains of *Lactobacillus casei* are capable of growing anaerobically on ribitol or xylitol by utilizing pathways unique to these organisms[56]. As noted earlier, the pentitols are transported into the cell by a substrate-specific PEP-phosphotransferase system that converts them to the corresponding pentitol phosphates. The latter are then converted to pentulose phosphate by NAD-specific ribitol phosphate or xylitol phosphate dehydrogenases. The pentulose phosphates are then converted to acetyl phosphate and glyceraldehyde-3 phosphate by a phosphoketolase[57]. Eventual products are ethanol, acetate, and a mixture of D- and L-lactate. The initial enzymes of the pathway–i.e., pentitol: phosphoenolpyruvate phosphotransferase, and pentitol phosphate dehydrogenase–are not stringently regulated by glucose or by intermediate products of glycolysis, but addition of glucose to a culture actively utilizing xylitol transiently represses pentitol phosphate dehydrogenase[111].

Production of pentitols is essentially a function of the NADPH-specific aldose reductase. Gong et al.[113] have described mutants of *C. tropicalis* that produce xylitol almost quantitatively from D-xylose. Microbiological production of polyols has been reviewed recently[11]. Earlier reviews considered the catabolism of polyols by yeasts[94] and other physiological aspects of polyol metabolism[95,96].

2.2.3 Phosphorylation

D-Xylulokinase catalyzes the phosphorylation of D-xylulose to D-xylulose-5-phosphate. This enzyme was first described in lactobacilli by Mitsuhashi and Lampen[114] and later purified from *Lactobacillus* by Stumpf and Horecker[115]. Wilson and Mortlock[85] found that two separate D-xylulokinases are inducible by D-xylose and D-arabitol in *Aerobacter aerogenes*. Both kinases have similar K_m values and substrate specificities. The latter D-xylulokinase from the D-arabitol operon has been purified to homogeneity and partially sequenced by Neuberger et al.[116]. It is active as a dimer with a subunit mol. wt. of 54,000. The apparent K_m for D-xylulose is 0.8 mM and the apparent V_{max} is 150 μmol min^{-1} mg^{-1} protein.

The presence of D-xylulokinase in yeasts is implied by the abilities of many yeasts to utilize D-xylulose under aerobic or anaerobic conditions[1-5,8-10,117]; however, an extensive computer-assisted literature search by the author did not uncover any papers pertinent to D-xylulokinase in yeasts of fungi published in the last 10 years. D-xylulokinase could be an important regulatory step in the metabolism of D-xylose.

2.3 The Pentose Phosphate Pathway

The PPP occurs widely in living cells. Its primary functions are to provide NADPH for biosynthetic reactions and ribose-5-phosphate for nucleotide synthesis[25]. The pathway consists of an oxidative phase that converts hexose phosphates to pentose phosphates, and a non-oxidative phase that converts pentose phosphates back to hexose phosphates.

The non-oxidative metabolic steps between D-xylulose-5-phosphate and D-fructose-6-phosphate occur essentially at equilibrium and are freely reversible, generating a pool of 3- to 7-carbon sugar phosphates[118]. Entry into the PPP occurs by several routes. With D-xylose as the carbon source, D-xylulose-5-phosphate is formed by the reactions described above; with D-glucose as a carbon source, or when the oxidative PPP is operating as a closed cycle as in *C. utilis*, D-ribulose-5-phosphate is generated by the decarboxylation of 6-phosphogluconate (Fig. 5). In most organisms, a non-oxidative route from D-glucose to pentose phosphate is

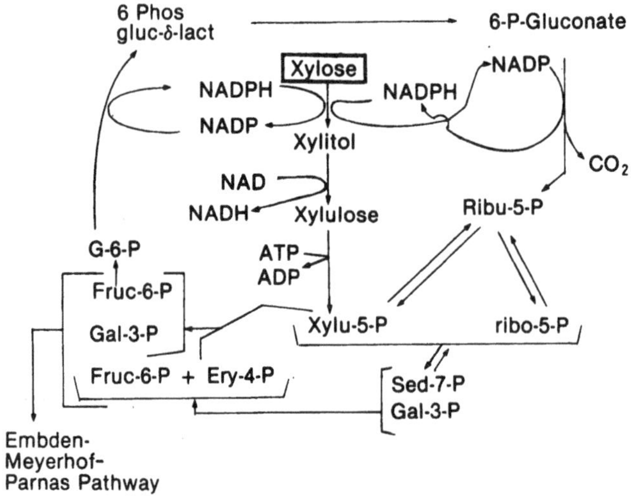

Fig. 5. Oxidative pentose phosphate pathway for the assimilation of D-xylose[97]. (M 151670)

employed. In this latter pathway, D-fructose-6-phosphate is converted to D-xylulose-5-phosphate and D-ribose-5-phosphate through the actions of transketolase [119-122], transaldolase [123], ribose phosphate isomerase, and ribulose phosphate-3-epimerase [124]. Carbon exits from the sugar phosphate pool by at least three and sometimes four routes: D-fructose-6-phosphate and glyceraldehyde-3-phosphate can enter the Embden-Meyerhof-Parnas pathways; D-ribose-5-phosphate is used for nucleotide synthesis; D-erythrose-4-phosphate is the starting point for the shikimic acid pathway leading to the synthesis of aromatic amino acids; and, in some organisms, D-xylulose-5-phosphate can form glyceraldehyde-3-phosphate plus acetyl phosphate through the action of phosphoketolase.

The basic elements of the non-oxidative PPP have been established for some time and have been the subjects of several extensive reviews [25,48,49,118,125]. New reaction sequences for the non-oxidative PPP in liver have recently been described [126-128]. Recent studies have also shown that a novel pentose phosphate cycle exists for formaldehyde fixation in methanol-utilizing yeasts such as *Hansenula* and *Candida*. One of the key enzymes of this cycle is the condensation of formaldehyde with D-xylulose-5-phosphate to give dihydroxyacetone and glyceraldehyde-3-phosphate in a transketolase-like reaction [129].

In certain yeasts, most notably *Candida* and *Rhodotorula* [130,131], the PPP is very active. *Candida*, for example, produces virtually all of its D-ribose from D-glucose through the oxidative PPP [130]. In *R. gracilis*, about 80% of the D-glucose is utilized via the non-oxidative and 20% by the oxidative reactions of the PPP [131]. Further degradation of D-xylulose-5-phosphate has been postulated to occur by way of phosphoketolase [132]. The oxidative PPP can provide two moles of NADPH for each mole of CO_2 released by this route (Fig. 5). In the absence of xylose isomerase, NADPH is required for the assimilation of D-xylose via D-xylose reductase; thus, at a minimum, one-tenth of the carbon would be oxidized to CO_2 to provide reducing power. In the process, NADH is generated through the activity of xylitol dehydrogenase [97].

2.4 Phosphoketolase

Phosphoketolase carries out the cleavage of either D-xylulose-5-phosphate or D-fructose-6-phosphate [133] to form glyceraldehyde-3-phosphate or D-erythrose-4-phosphate plus acetyl phosphate:

$$\text{D-xylulose-5-phosphate} + \text{phosphate} \xrightarrow[\text{TPP}]{\text{Mg}^{++}} \text{glyceraldehyde-3-phosphate} + \text{acetyl phosphate} + H_2O \qquad (2)$$

The overall free energy change for this reaction has been estimated to be —10.5 kcal mol^{-1} [134]. If the acetyl phosphate is in turn linked to the production of ATP and acetate [135], the free energy change is an additional —3.1 kcal mol^{-1} [134]:

$$\text{Acetyl phosphate} + \text{ADP} \rightarrow \text{acetate} + \text{ATP} \qquad (3)$$

In either case, the reaction represents a substantial driving force for pentose utilization.

The role of phosphoketolase in the fermentation of pentoses by *Lactobacillus* was described by Fred et al. in 1921 [136]. They showed that both D-xylose and L-arabinose were converted quantitatively to equimolar amounts of acetic and lactic acids, and they postulated that a cleavage to 2- and 3-carbon fragments was involved. Subsequent isotope studies [137–139] supported this hypothesis, and the existence of phosphoketolase in *Lactobacillus* was eventually demonstrated [135]. Based on isotope studies, Gibbs et al. [140] postulated a similar mechanism for the fermentation of D-xylose in *Fusarium lini* as early as 1954, but to the author's knowledge, phosphoketolase has never been demonstrated in this fungus.

Phosphoketolase was first believed to be unique to the lactobacilli and *Leuconostoc* [141,142]; but more recent studies have shown its existence in *Pediococcus pentosaceus* [143–144], *Bifidobacterium bifidum (Lactobacillus bifidus)* [134,146], *Bacterioides ruminicola* [147], *Thiobacillus* [148,149], and in several yeasts including *Rhodotorula graminis*, *Rhodotorula glutinis*, *Candida tropicalis*, *Candida humicola*, and *Candia* 107 [133,150,151]. In most instances the levels of phosphoketolase are low, and the physiological significance in some organisms has been questioned [133,147–149]. But the enzyme is inducible by pentoses [143] and gluconate [145], and it might be important during anaerobic growth on D-xylose, because reduction of the two-carbon moiety to ethanol would provide an alternative pathway for the regeneration of NAD from NADH.

The biochemical mechanism of phosphoketolase requires Mg^{2+} and thiamine pyrophosphate, and involves the formation of an "active glycolaldehyde" intermediate [152]. The initial cleavage of D-xylulose-5-phosphate involves the formation of dihydroxyethyl thiamine diphosphate [153] which is the precursor of acetyl phosphate. Although the initial cleavage is reversible, the step involving formation of an acetyl group from the dihydroxyethyl group is irreversible, thus accounting for the irreversibility of the entire reaction [154]. Depending on the reaction conditions, different products are formed. In the presence of phosphate, phosphoketolase converts glycolaldehyde to acetyl phosphate; and in the presence of ferricyanide, glycolaldehyde is converted to glycolate [152]. The activity of phosphoketolase is markedly inhibited by NADH, NADPH, ATP, and acetyl-CoA, and is reduced in the presence of transketolase, which competes with it for the same substrate [133]. In the presence of ferricyanide, transketolase also catalyzes the oxidative cleavage of D-xylulose-5-phosphate or D-fructose-6-phosphate into glycolate and glyceraldehyde 3-phosphate of D-erythrose-4-phosphate, respectively [155], suggesting that the phosphoketolase and transketolase reactions may be very similar.

3 Regulation of D-Xylose Metabolism

Classical studies of metabolic regulation in yeasts have examined the changes that occur when cells are shifted between aerobic and anaerobic conditions when they undergo a transition from starvation to growth or when they adapt to a new set of nutritional factors. Because anaerobic metabolism of pentoses was unknown in yeasts until recently, studies of anaerobic pentose conversions by yeasts are preliminary and incomplete. This section attempts to review several well-studied regulatory phenomena of hexose metabolism and compare then with mechanisms in pentose metabolism. Regulation of hexose metabolism in yeasts has been recently reviewed [51].

3.1 Aerobic and Anaerobic Utilization of D-Xylose

Even though roughly half of all yeasts are capable of assimilating D-xylose under aerobic conditions, until recently none were known to metabolize this sugar or other pentoses anaerobically [77]. Anaerobic conversion of D-xylose to ethanol is now recognized in the yeast *Pachysolen tannophilus*, and many yeasts have been shown to convert the keto isomer of D-xylose, D-xylulose, to ethanol under anaerobic conditions (see below). However, anaerobic growth of yeasts (including *P. tannophilus*) on D-xylose has not been reported. In most cases, aeration stimulates ethanol production from D-xylose [13,14,17,18], and in *Candida tropicalis* aeration has been reported as essential for ethanol production [16]. Aeration is also required for the optimal production of 2,3-butanediol by *Klebsiella pneumoniae* [156] (see also this volume, p. 101). The biochemical or physiological bases for these requirements for oxygen in the metabolism of D-xylose are not yet fully understood, but they could be related either to pentose transport, to the reductive/oxidative pathway for D-xylose assimilation, to the regeneration of reduced coenzymes or to the production of ATP for growth through oxidative phosphorylation.

In discussing aerobic and anaerobic metabolism it is convenient to use the term "fermentation" in its classical sense — i.e., the utilization of an organic compound as both a donor and a recipient of electrons. During fermentation, the substrate is cleaved or otherwise metabolized with part of it going to more oxidized products (such as CO_2) and part going to more reduced products (such as ethanol). Energy for metabolism is derived from substrate-level phosphorylation and not through respiratory processes with oxygen (or other exogenous compounds such as nitrate or sulfate) as the terminal electron acceptor. Because the substrate is not fully oxidized to CO_2 and water during fermentation, only a fraction of the chemical energy stored in the substrate is recovered as ATP for metabolic processes. Aerobic respiratory metabolism is much more efficient in the production of ATP than is fermentation. On the other hand, fermentation is more efficient in conserving chemical energy in metabolic end products such as ethanol. Yeast cells exercise metabolic regulation in utilizing growth substrates via aerobic or anaerobic pathways.

Pasteur was the first to demonstrate that a *Saccharomyces* yeast growing on low concentrations of D-glucose decreases its D-glucose uptake when subjected to aeration [157]. Respiration of D-glucose to CO_2 and water greatly increases the yield of ATP per mole of sugar metabolized over that obtained from the production of ethanol. The resultant increase in the ATP/ADP ratio and the increased concentration of certain other metabolites, most notably citric acid, feed back to inhibit the activity of phosphofructokinase [50]. The intracellular concentration of D-glucose-6-phosphate thereby increases and it in turn inhibits D-glucose transport. This phenomenon, known as the Pasteur effect, is generally considered one of the most basic regulatory mechanisms in yeasts. It should be noted, however, that more recent interpretations by Lagunas [158] have indicated that the Pasteur effect is only marginally expressed in *Saccharomyces*.

Saccharomyces cerevisiae, along with *Saccharomyces carlsbergensis*, *Saccharomyces lactis*, and *Schizosaccharomyces pombe* are considered "glucose sensitive yeasts" or "Gärungshefen" [51,159]. With these organisms, expression of respiration is dependent on the absence of even minor concentrations of glucose. The absence of oxygen diminishes cell yield, abolishes respiration, and stimulates uptake and glycolytic utilization of glucose. In contrast, the "Atmungshefen" are insensitive to glucose repression of respiration. These yeasts include *Candida utilis* and *Hansenula* [51,159].

Aeration stimulates sugar utilization in some yeasts. With the genus *Brettanomyces*, the alcoholic fermentation of glucose is inhibited under anaerobic conditions. This phenomenon is known as the negative Pasteur effect or Custers effect [160-165]. A similar, but lesser, stimulation of CO_2 production by oxygen during the fermentation of glucose has been observed with the genera *Pachysolen*, *Pichia*, *Endomycopsis*, and *Kluyveromyces* [163]. Strictly aerobic yeasts such as *Rhodotorula* [52,81] are also recognized. With these yeasts no sugar utilization occurs under anaerobic conditions. *Rhodotorula* uses the PPP almost exlusively [132]. In a wide variety of yeast species, but particularly those belonging to *Candida*, *Hansenula*, *Kluyveromyces*, *Metschnikowia*, *Pachysolen*, *Pichia*, and *Torulopsis*, oxygen is required for the utilization of one or more sugars. This is commonly known as the Kluyver effect [166].

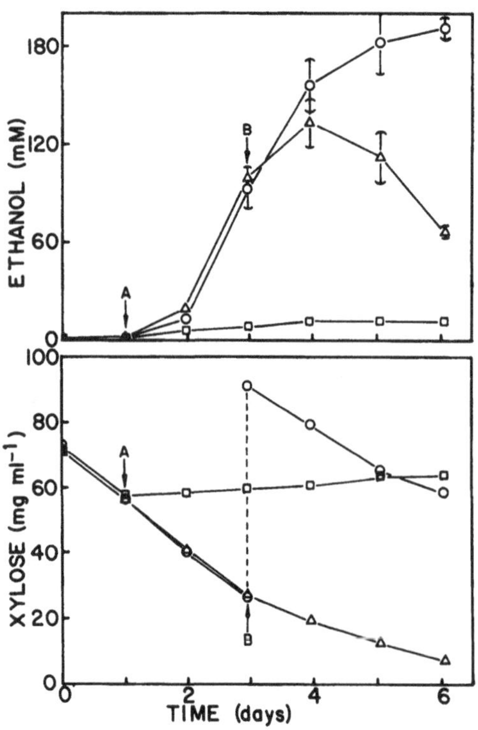

Fig. 6. Ethanol production by *Candida tropicals* ATCC 1369 under aerobic conditions. Cells were grown in 33 ml yeast nitrogen base (Difco) plus 7.5% D-xylose (YNX-7.5%) at 28 C with shaking at 200 rpm; initial cell growth was rapid and preceded the appearance of ethanol.
After 1 d (A), some cultures were switched to anaerobic conditions by replacing cotton plugs with rubber stoppers and flushing with nitrogen (□-□). After 3 d (B), other cultures were supplemented with additional xylose (○-○). Control cultures (△-△) were maintained under original conditions. Brackets show standard deviations obtained in triplicate flasks. (M 149 602)

Initial studies indicate that the Pasteur effect is not manifested during the metabolism of D-xylose by *C. tropicalis*. To the contrary, a negative Pasteur effect is indicated as D-xylose consumption and ethanol production stop when cells of *C. tropicalis* are shifted from aerobic to anaerobic conditions (Fig. 6). In this organism ethanol production from D-xylose is also prevented by the addition of respiratory inhibitors such as sodium azide or antimycin A, but not by the tricarboxylic acid cycle inhibitors fluoroacetate or fluorocitrate (Fig. 7). These findings

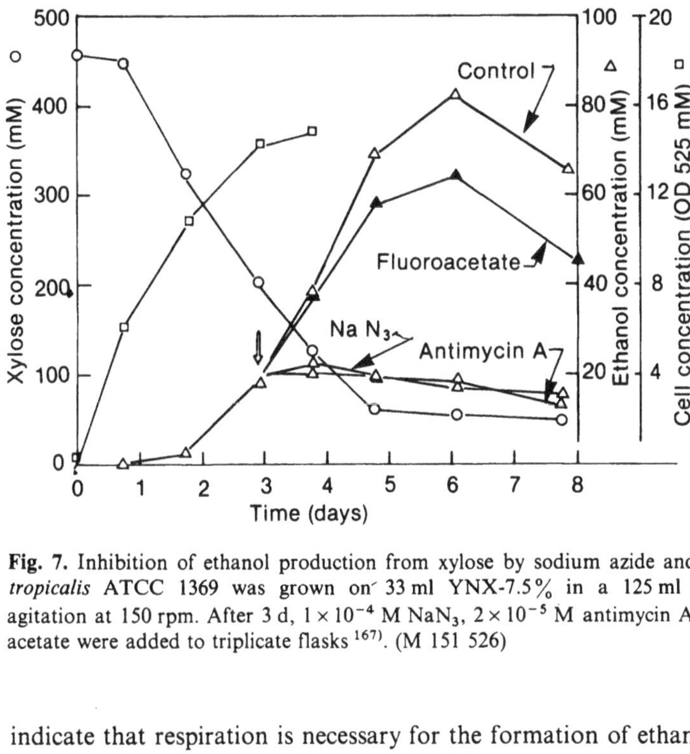

Fig. 7. Inhibition of ethanol production from xylose by sodium azide and antimycin A. *Candida tropicalis* ATCC 1369 was grown on 33 ml YNX-7.5% in a 125 ml Erlenmeyer flask with agitation at 150 rpm. After 3 d, 1×10^{-4} M NaN_3, 2×10^{-5} M antimycin A, or 8×10^{-4} M fluoroacetate were added to triplicate flasks [167]. (M 151 526)

indicate that respiration is necessary for the formation of ethanol from D-xylose by *C. tropicalis*. The metabolic basis for this requirement could be attributable to either the necessity for respiratory regeneration of NAD from NADH formed in the assimilation of D-xylose, or to the existence of an active transport system for D-xylose [167]. Present evidence does not allow differentiation between these alternatives. Other researchers have reported that various *Candida* mutants are capable of converting D-xylose to ethanol under "fermentative" conditions [11,17,18]. These experiments, however, were not strictly anaerobic because a small-gauge needle was used to vent the cultures, which were incubated with agitation. Moreover, aeration stimulated the rate of ethanol production, indicating that oxygen might have been limiting the rate of conversion.

In contrast to *C. tropicalis*, *P. tannophilus* produces ethanol from D-xylose under strictly anaerobic conditions [13,167]. When sealed vials were flushed with N_2, pregrown cells of *C. tropicalis* produced no ethanol from D-xylose, but *P. tannophilus* produced a mixture of ethanol and acetic acid (Fig. 8). The presence of this latter

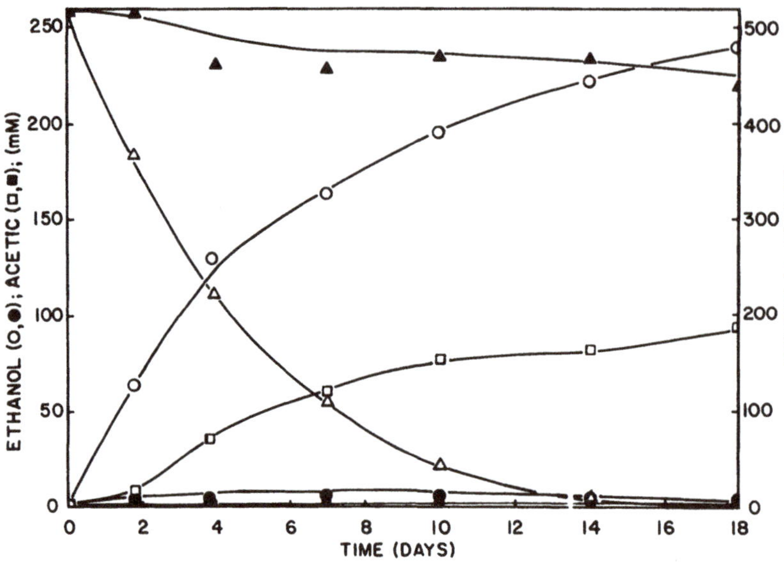

Fig. 8. Comparison of *Pachysolen tannophilus* NRRL Y-2460 and *Candida tropicalis* ATCC 1369 for ethanol production under anaerobic conditions. Cells of both organisms were grown in 33 ml of YNX-7.5% with agitation at 150 rpm for 3 d harvested by centrifugation and inoculated to 33 ml of the same medium in a 50 ml serum vial. Vials were sealed with thick butyl rubber bungs (Belco) and sparged with N_2. Inoculum per vial: *P. tannophilus* = 0.17 mg dry weight (open symbols), *C. tropicalis* = 0.07 mg dry weight (closed symbols) [167]. (M 151 603)

product suggests that a phosphoketolase pathway might be present in *P. tannophilus*. Even though aeration is not required for ethanol production by *P. tannophilus*, a low aeration rate stimulates ethanol production [167].

3.2 D-Glucose-6-Phosphate Dehydrogenase

Regulation of the oxidative PPP occurs largely through regulation of D-glucose-6-phosphate dehydrogenase activity [168,169]. This enzyme catalyzes the oxidation of D-glucose-6-phosphate to 6-phosphoglucono-δ-lactone, which is subsequently hydrolyzed to 6-phosphogluconate by lactonase:

$$\text{D-glucose-6-phosphate} + \text{NADP}^+ \rightarrow \text{6-phosphoglucono-}\delta\text{-lactone} + \text{NADPH} + \text{H}^+ \quad (4)$$

$$\text{6-phosphoglucono-}\delta\text{-lactone} + \text{H}_2\text{O} \rightarrow \text{6-phosphogluconate} \quad (5)$$

Because of the large free-energy change of these two reactions ($\Delta G^{\circ\prime} = -5.1$ kcal mol^{-1} at 25 C, pH 7.0), the step is essentially irreversible, and would be expected to go virtually to completion. However, this is prevented by the fact that D-glucose-6-phosphate dehydrogenase is strongly inhibited by one of its products, NADPH [170].

This inhibition is competitive with NADP, so, by reversible inactivation of the enzyme, a steady state can be established that is rather far removed from the thermodynamic equilibrium. In rat livers, this difference amounts to a factor of about 10^8, and control of enzyme activity is essentially a matter of de-inhibition [171]. The D-glucose-6-phosphate dehydrogenases of *Saccharomyces carlsbergensis* [172] and *S. cerevisiae* [173] are strongly inhibited by NADPH, but no studies have shown that it regulates these enzymes [174]. It has been reported that the NADPH/NADP ratio in aerobically grown *S. cerevisiae* is near unity [175], so under these conditions in this organism, the D-glucose-6-phosphate dehydrogenase may be (essentially) de-inhibited.

Bacterial D-glucose-6-phosphate dehydrogenases do not appear to be regulated in a manner equivalent to that found in yeasts and rat livers. *Leuconostoc mesenteroides* possesses a single D-glucose-6-phosphate dehydrogenase which is active with both NAD and NADP [176]. The enzyme does not have a high affinity for either of these coenzymes [177], and it has been suggested that they do not play a significant role in its regulation [176]. In *Acetobacter xylinum*, two D-glucose-6-phosphate dehydrogenases are present, each with specificity for either NAD or NADP. Although the reduced coenzymes inhibit activity of their respective enzymes, no regulatory role has been proposed [178].

Other reported metabolic inhibitors of D-glucose-6-phosphate dehydrogenase are ATP [179], acetyl-CoA [180], erythrose-4-phosphate and glyceraldehyde-3-phosphate [181]. The significance of these compounds *in vivo* may be questionable in some organisms. Biochemical and other regulatory aspects of D-glucose-6-phosphate dehydrogenase have been reviewed recently [174].

3.3 Nutritional Factors

After carbon source and aeration, the source of nitrogen supplied to a culture is probably next in importance in determining how the cell regulates its metabolic machinery. In yeasts, ammonium ions counteract the inhibition of phosphofructokinase by ATP; therefore NH_4^+ stimulates glycolysis [50]. The PPP is also regulated by nitrogen. Ammonium salts have been found to stimulate the oxidative PPP in *S. cerevisiae* [182]. Because D-glucose-6-phosphate is normally inhibited by NADPH and because incorporation of NH_4^+ into α-ketoglutarate requires NADPH, the addition of ammonium salts stimulates growth, decreases the intracellular level of NADPH, derepresses D-glucose-6-phosphate dehydrogenase, and thereby increases the activity of the oxidative PPP.

Onishi has shown that polyol formation by *Pichia miso* is profoundly affected by the C/N ratio [183,184]. Higher levels of polyols are observed in low-nitrogen than in a high-nitrogen medium. Polyols would accumulate if NADPH production increased relative to the rate of NAD regeneration. Such a situation would occur with a stimulation of the oxidative PPP and a repression of respiration.

The form of nitrogen also affects activity of the PPP. Osmond and Rees [185] have reported that when *Candida utilis* is grown on nitrate, the levels of D-glucose-6-phosphate dehydrogenase and transketolase are about 2.5 times higher than when cells are grown on a complex medium that contains amino acids. Presumably, this shift occurs in response to a greater demand for NADPH in the nitrate-grown cells. Similar observations have been made with *Aspergillus nidulans*[186] and with sycamore cells grown in tissue culture[187].

4 Utilization of D-Xylose, D-Xylulose, and Xylitol by Yeasts and Fungi

As was noted earlier, roughly half of all yeast strains tested will utilize D-xylose for growth under aerobic, but not anaerobic, conditions[77]. However, many organisms will convert D-xylose to xylitol, even though they are not able to grow on the pentose. Once D-xylose is isomerized to D-xylulose, many yeasts will use it both oxidatively and fermentatively. Xylitol is generally a poor carbon source for yeasts, even though it is the putative intermediate in D-xylose assimilation.

4.1 D-Xylose

The utilization of D-xylose is not limited to those yeasts able to grow on it oxidatively. Some yeasts will convert it to xylitol or ethanol even though they do not grow on it. Gong et al.[188] tested 20 strains of *Candida* belonging to 11 species, 21 strains of *Saccharomyces* belonging to 8 species, and 8 strains of *Schizosaccharomyces pombe* for their abilities to utilize D-xylose. All of the *Candida* strains grew on D-xylose; the *Saccharomyces* and *Schizosaccharomyces* strains grew on D-xylose poorly or not at all and, hence, D-glucose was used to grow these cells. Pregrown cells were then tested for their abilities to utlize D-xylose under high or low aeration. All of the *Candida* species consumed D-xylose and produced xylitol in moderate yield (10–50%). Arabitol was the second most common product. Some *Candida* species also produced ethanol in relatively low yield (less than 10%). Species producing ethanol from D-xylose included *Candida blankii*, *Candida friedrichii*, *Candida lusitaniae*, *Candida parapsilosis*, *Candida pseudotropicalis*, *Candida solani*, *Candida steatolytica*, and *Candida tropicalis*. The *Saccharomyces* species consumed much less D-xylose, but produced xylitol in roughly the same yields. Nine strains of *Saccharomyces cerevisiae* produced xylitol in concentrations ranging from 0.1% to 0.7%. Strains of *S. pombe* consumed relatively more D-xylose than *Saccharomyces*, but made less xylitol. *Schizosaccharomyces pombe* also produced ethanol from D-xylose in concentrations ranging from 0.1% to 0.5%.

Maleszka and Schneider[189] screened 15 yeasts capable of assimilating D-xylose for their abilities to convert D-xylose, D-xylulose, and xylitol to ethanol under low aeration conditions (semi-aerobic). D-xylose utilization was tested under both aerobic and anaerobic conditions, and with rich undefined or defined media. In almost all instances, ethanol production was better on rich media under semi-aerobic conditions. All yeasts tested produced some ethanol from D-xylose. The highest ethanol concentration (1%) was obtained with *Pachysolen tannophilus*, followed by *Candida guilliermondii*, *Candida terebra*, and *Pichia guilliermondii*.

Slininger et al.[15] determined the optimum pH and the specific rate of ethanol production from D-xylose by *P. tannophilus* in a 10-liter bioreactor. Both alcohol production and growth were optimal when the incubation temperature was 32 C, when the pH was near 2.5, and when D-xylose and ethanol concentrations did not exceed 50 g l^{-1} and 20 g l^{-1}, respectively. Yeast cell growth occurred at a maximum specific rate of 0.18 h^{-1} during the logarithmic growth phase. By comparison, this is a little less than half the rate attained by brewer's yeast on glucose. The maximum specific ethanol production rate was 0.08 g ethanol g^{-1} cells h^{-1}, and the maximum rate of D-xylose consumption was 0.3 g g^{-1} h^{-1}. The maximum yield attained was 0.34 g ethanol g^{-1} D-xylose consumed. The aeration rate was controlled at 0.075 min^{-1} (volume air/volume medium) with a stirring rate of 100 rpm, but it is not known whether or not this represents an optimum. Aeration was required for growth but not for ethanol production, and the specific rate of ethanol production was found to be the same for aerated and unaerated cells.

The ability of *Fusarium lini* to ferment pentoses has been known for 60 years [190], and has recently received additional study. White and Willaman [19] demonstrated that *F. lini* could form up to 4.3% ethanol from D-glucose, but somewhat less from D-xylose. The fermentation rate with D-xylose, however, was extremely slow. Up to 40 days were required to consume a 4% solution of D-xylose. More recent studies have shown that the rate of fermentation and ethanol yield are dependent upon the *Fusarium* sp. and the strain tested. Suihko and Enari [23] recently surveyed 26 strains of *Fusarium* representing 12 species for their abilities to ferment 5% solutions of D-glucose or D-xylose. Various strains formed between 0.4% and 2.1% ethanol from D-xylose in 6 days. The best strain tested was *Fusarium oxysporum* VTT-D-80134. It formed 2.5% ethanol from 5% D-glucose in 2.5 days and from 5% D-xylose in 6 days.

Immobilization of *Mucor* sp. and *F. lini* has been attempted as a means to increase the volumetric rate of ethanol production. Immobilized *F. lini* formed up to 1% ethanol within 24 h, but *Mucor* formed relatively little ethanol under the conditions employed [191].

4.2 D-Xylulose

The pattern of D-xylulose utilization differs from that for the utilization of D-xylose. Gong et al. [188] found that most yeasts utilized D-xylulose readily under aerobic or semi-aerobic conditions. With *Candida* species, ethanol concentrations from D-xylulose were generally higher and xylitol concentrations generally lower than when D-xylose was used as the carbon source. *C. tropicalis* ATCC 20240 formed 2% ethanol plus 0.2% xylitol and 0.2% arabitol from 4.67% D-xylulose. Of the *Candida* strains tested, only *C. parapsilosis* ATCC 28474 formed more xylitol than ethanol from D-xylulose. This organism also formed large amounts of xylitol from D-xylose. The pattern of D-xylulose utilization by *Saccharomyces* strains was similar to that for *Candida*, but with relatively less ethanol and relatively more xylitol being formed. Arabitol production was appreciably higher with *Saccharomyces* than with *Candida* species. All strains of *S. pombe* tested produced ethanol from D-xylulose in concentrations ranging between 1.2% and 2.3%. Xylitol and arabitol were also produced in concentrations ranging from 0.1% to 1.1%, and 0.1% to 0.3%, respectively. *Schizosaccharomyces pombe* ATCC 2476 produced the most ethanol (2.3%) and least xylitol (0.1%) from D-xylulose of all the yeast strains tested.

Maleszka and Schneider [189] found that yeasts that utilized D-xylose readily for the production of ethanol utilized D-xylulose relatively less. This was particularly true of *P. tannophilus* ATCC 2460. Gong et al. [188] likewise found that *C. blankii* ATCC 18735 produced more ethanol from D-xylose than any other *Candida* species tested, but did not form ethanol from D-xylulose. Both of these organisms will readily assimilate D-xylose. It should be noted that *S. pombe* strains do not assimilate D-xylose, but will readily use D-xylulose aerobically or anaerobically. *Fusarium oxysporum* f. sp. *lini* and *Aspergillus niger* preferentially assimilate D-xylose in the presence of D-xylulose. This physiological trait can be employed to purify D-xylulose [192].

4.3 Xylitol

Even though xylitol is formed in relatively large amounts from D-xylose and D-xylulose, and even though it is believed to be an intermediate in the assimilation of D-xylose by yeasts, no yeast has been reported to utilize xylitol efficiently for the production of ethanol. Gong et al. [188] found that *C. tropicalis*, *C. pseudotropicalis*, and *Kluyveromyces fragilis* formed some ethanol (0.2% to 0.5%) from xylitol. Maleszka and Schneider [189] reported only trace amounts of ethanol produced from the 15 yeast strains tested. Barnett [94] found that *Torulopsis candida* was the most versatile of 16 strains of yeasts tested for their abilities to utilize polyols. This organism, however, is not active anaerobically [193]. *Candida utilis* did not use any of the polyols tested, even though it possesses enzymes for the NAD-linked oxidation of xylitol to D-xylulose [94].

Both *Mucor* sp. and *Fusarium* have been shown to ferment xylitol to ethanol. One strain of *Mucor* yielded 0.41 g ethanol g^{-1} xylitol consumed, attaining a concentration of 0.9% ethanol in 4 days. Rates and yields with *Fusarium* were somewhat lower [194].

4.4 Sugar Mixtures

The utilization of mixtures of D-xylose with D-glucose and L-arabinose is of particular significance, as these sugars are all present in hydrolysates of agricultural and forestry residues. Hsiao et al. [195] compared the abilities of four yeasts (*S. cerevisiae*, *S. pombe*, *C. utilis*, and *R. toruloides*) to utilize D-glucose, D-xylose, D-xylulose, and xylitol. The sugars were tested separately and in various mixtures. When D-glucose was present, utilization of D-xylose, D-xylulose, and xylitol was generally inhibited. In the absence of D-glucose, however, the pentoses and pentitol were consumed simultaneously. Maleszka et al. [196] have investigated the abilities of various yeasts, especially *Candida lusitaniae* to utilize D-cellobiose in the presence of D-xylose. These researchers found that *C. lusitaniae* produced more ethanol from a mixture of D-xylose and D-cellobiose than was produced when the two sugars were fermented separately; that is, a synergism seemed to exist such that the D-cellobiose stimulated the production of ethanol from D-xylose or vice versa.

The ability to utilize a mixture of D-xylulose and D-xylose anaerobically appears to be widespread, but not universal, among the yeasts. In a survey of 32 yeasts and 3 strains of the bacterium *Zymomonas mobilis*, Jeffries [10] found that about 40% of the yeast strains were capable of forming ethanol from a mixture of D-xylulose and D-xylose under anaerobic conditions. Two species each of *C. tropicalis* and *S. pombe* formed ethanol rapidly and without an initial lag. Six strains of *S. cerevisiae* that utilized D-glucose rapidly in the presence of D-xylose formed ethanol from a mixture of D-xylulose and D-xylose at a slower rate. Generally, the rate of pentose utilization was about one tenth that for D-glucose. Ten yeasts capable of utilizing D-glucose did not form significant amounts of ethanol from the pentose mixture. Particularly notable was the inability of three strains of *Z. mobilis* to utilize D-xylulose, even though they utilized D-glucose rapidly. This inability has been attributed to the fact that *Zymomonas* apparently lacks transketolase and transaldolase [197].

4.5 Hydrolysates of Hemicellulose

Acid hydrolysates of hemicellulose are substantially more difficult for microbes to utilize than are corresponding mixtures of pure sugars. This recalcitrance is attributable largely to the presence of sugar-degradation products, particularly furfural. However, furfural toxicity does not seem to account completely for the observed inhibition, and other unidentified compounds seem to be present as well. These inhibitory factors necessitate a purification of the hydrolysate prior to microbial utilization. Acid hydrolysates of hemicellulose can be neutralized by calcium oxide to precipitate sulfuric acid as $CaSO_4$ [12,18,46]. This procedure also removes phosphate, which must be replaced. Ion-exchange chromatography and treatment with activated carbon are also effective for improving the utilization of acid hydrolysates by yeasts [46].

Gong, Ladisch, and Tsao [18] have reported the conversion of an acid hydrolysate of wood to ethanol using a mutant *Candida* sp. XF 217. The direct conversion of sugars (predominantly D-xylose) to ethanol occurred more rapidly under aerobic than under semi-aerobic conditions, and the rate of semi-aerobic conversion was markedly enhanced by the addition of D-xylose isomerase to the medium. Approximately 2% ethanol was attained under all three conditions. This value represented an ethanol yield of about 90% of theoretical based on the sugars originally present. Hsiao et al. [12] have obtained similar results with the fermentation of acid hydrolysates of corn stover by *S. cerevisiae* using D-xylose isomerase and borate to enhance the formation of D-xylulose. It should be noted, however, that L-arabinose, which is present in significant amounts in corn stover, is not utilized by this process.

Even though both *Mucor* and *Fusarium* sp. will ferment D-glucose and D-xylose, only *Mucor* will readily ferment hemicellulose hydrolysates; *Fusarium* appears to be particularly sensitive to toxic compounds present in the acid hydrolysate. Starting with a hemicellulose hydrolysate containing 4.4% D-xylose, 1.4% D-glucose, and 0.9% L-arabinose, *Mucor* sp. 105 formed 2% ethanol and 0.9% xylitol within 3 to 5 days [194].

5 Depolymerization and Fermentation

Bacteria, yeasts, and fungi have all been reported to possess xylanases and the necessary enzymes for debranching and removing substituents [198,199]. Relatively few organisms, however, have the abilities to both degrade xylan and ferment the resultant D-xylose. To get around this limitation, two or more organisms have been used in coculture. The simultaneous saccharification and fermentation of xylans and cellulose has been demonstrated with both bacterial and fungal systems [200–203].

The most extensively studied bacterial system employs *Clostridium thermocellum* in coculture with either *Clostridium thermosaccharolyticum* [58,59,61] or *Clostridium thermohydrosulfuricum*. The *C. thermocellum* cellulase exhibits xylanase activity, but *C. thermocellum* does not utilize D-xylose. In the absence of a second organism, reducing sugars accumulate in the medium. Either *C. thermosaccharolyticum* or *C. thermohydrosulfuricum* will ferment D-xylose to ethanol plus acetic and lactic acids. *Clostridium thermoaceticum* will also ferment D-xylose with acetate as the only product [204,205].

The native strains of these clostridia generally exhibit relatively low tolerance to ethanol, but strain selection programs have resulted in the recovery of isolates of *C. thermosaccharolyticum* that will form up to 2.7% ethanol from D-xylose with a yield of 0.35 g ethanol g^{-1} D-xylose consumed. In addition, cultures form 0.7% acetic and 0.25% lactic acids within 48 h. In coculture, selected strains of *C. thermocellum* and *C. thermosaccharolyticum* will form 2.5% ethanol from cellulose in about 96 h. The yield of ethanol from Wiley-milled corn stover, however, is lower; cocultures form about 1% ethanol and 0.6% acetic acid in 120 h [200].

Xylan-degrading activity has been reported in about 25% of yeasts and yeast-like organisms capable of thriving on D-xylose. Of 54 strains tested, 13 belonging to the genera *Aureobasidium*, *Cryptococcus*, and *Trichosporon* grow on xylan. None of these strains, however, are fermentative; they utilize D-xylose oxidatively [199].

Numerous fungi are known to degrade both cellulose and hemicellulose, but only a few, especially *Fusarium* spp. [19–23,206–209] and *Monilia* [210], exhibit both depolymerase and fermentative activity. Although *Monilia* has been shown to produce both cellulase and xylanase and to ferment both D-glucose and D-xylose, the direct fermentation of polymer to ethanol has been demonstrated only with cellulose [210]. No direct fermentation of xylan to ethanol has yet been demonstrated with *Fusarium* or any other single organism. The direct microbiological conversion of cellulosics to ethanol has recently been reviewed [211].

6 Implications for Strain Selection and Process Design

It should be apparent from the foregoing that no available yeast or bacterium is completely satisfactory for the conversion of D-xylose to ethanol. Limiting factors include the conversion rate, product yield, and low ethanol tolerance. While these elements may be amenable to strain selection and contemporary recombinant DNA technology (e.g., see review by Schneider et al., this volume), it is likely that more basic knowledge about the physiology and biochemistry of D-xylose fermentation is necessary. The rate-limiting biochemical steps have been surmised but not yet identified. Transport, conversion to D-xylulose, and phosphorylation are all candidates. Assuming that the metabolism proceeds through fructose-6-phosphate and the Embden-Meyerhoff-Parnas pathway, phosphofructokinase could also be a regulatory, rate-limiting step. Considering the relative ease with which D-xylulose is utilized by yeasts, the incorporation and expression of D-xylose isomerase in an ethanol-tolerant yeast such as *S. cerevisiae* or *S. pombe* could significantly improve the rate of D-xylose fermentation. In this regard, a reexamination of the presence of D-xylose isomerase in *Candida* and *Rhodotorula* is in order. Reduced xylitol production is also essential. This trait apparently can be altered through conventional mutagenesis [17)], but complete eliminaton will probably require elimination of aldose reductase activty. The reported abilities of *S. cerevisiae* and *S. pombe* to produce xylitol even though they do not grow on D-xylose [188)] suggest that other nonspecific reductases may function in this regard as well. The physiological role of phosphoketolase, both as a catalyst for a potential driving reaction and as a source of acetic acid production, requires further research as do the physiological and biochemical bases for inhibition of anaerobic growth and stimulation of ethanol production by aeration. Attempts to improve tolerance to ethanol should include supplementation of media with ergosterol and other membrane components as well as selection of alcohol-resistant strains.

From a practical point of view, there are two basic approaches to the conversion of D-xylose to ethanol by yeasts. One is to use a two-stage system based on D-xylose isomerase and yeast; the other is to use a single-stage system with a selected yeast such as *P. tannophilus*. Each approach has advantages and disadvantages.

The two-stage system has undergone substantially more process development and is probably in a better position for commercialization. The characteristics of this approach have been reviewed more extensively in a previous volume in this series [11)]. In general, the two-stage system has the advantages of a higher overall rate of conversion and probably a higher ethanol yield as well [12)]. The organism of choice in a two-stage process appears to be *Schizosaccharomyces pombe* ATCC 2476 because it utilizes D-xylulose rapidly and produces relatively little xylitol [188)]. The two-stage process seems to accumulate a higher concentration of ethanol than does a single-stage conversion. About 4.6% (wt/wt) ethanol has been produced by *C. tropicalis* and 6.3% by *S. pombe* using two-stage systems [9,10)]. An additional mitigating factor in the two-stage system is the cost of D-xylose isomerase. If one assumes that the cost of isomerization is equivalent to the prevailing cost in the production of high-fructose corn syrup, it will add about US \$ 0.03 kg^{-1} (US \$ 0.015 lb^{-1}) to the cost of the D-xylulose fermented [10)]. It is possible, however, that the use of borate in the isomerization could substantially lower this expense [12)].

Pachysolen tannophilus appears to be the organism of choice for the direct, single-stage conversion. It is capable of using D-xylose under anaerobic conditions, thereby minimizing losses of ethanol through subsequent respiration. Slininger et al. [15)] have reported that *P. tannophilus* yields 0.34 g ethanol g^{-1} pentose consumed. This corresponds to about 66% of the theoretical yield. Acetic acid and xylitol are significant byproducts. Various *Candida* spp. and mutants have been reported as capable of converting D-xylose to ethanol [16,17)]. However, none of the published studies have described experiments beyond the flask scale, and formation of xylitol is significant with these organisms as well. In general, the direct conversions of D-xylose to ethanol proceed at about half the rate of the D-xylulose fermentation [194], and the rate of D-xylulose fermentation is slower than that for D-glucose [10]. These rates might be improved through process optimization, but it is likely that the slow rates for pentose fermentations result from the biochemical pathway employed. The alcohol tolerance of *P. tannophilus* has been reported as relatively low [15]. Immobilized cells of *P. tannophilus* will produce 1.5% to 2% ethanol on a continuous basis for several weeks, but the rates and yields are low [167].

Single-stage direct microbiological processes for converting polymeric materials to ethanol and other chemicals possess several favorable characteristics. Most importantly, the use of a single bioreactor simplifies the process and reduces the capital expenditure. Potentially, the overall rate of conversion can be increased because intermediate products are removed as they are formed, thereby relieving feedback inhibition and catabolite repression of depolymerase activity. In practice these advantages are not necessarily realized because the rate obtained under optimal conditions for the overall process might not be greater than the rate obtained through optimization of the individual process steps. Still, the simplicity inherent in a single-stage process tends to promote further research to that end.

7 Acknowledgements

The author wishes to thank D. E. Eveleigh, G. F. Leatham, and H. Schneider for critically reading the manuscript, and T. K. Kirk for advice and encouragement.

8 References

1. Wang, P. Y., Shopsis, C., Schneider, H.: Biochem. Biophys. Res. Comm. *94*, 248 (1980)
2. Schneider, H., et al.: Pentose fermentation by yeasts. In: Current Developments in Yeast Research. (Steward, G. G., Russell, I. eds.), p. 81, Toronto: Pergamon Press 1981
3. Wang, P. Y., Johnson, B. F., Schneider, H.: Biotechnol. Lett. *2*, 273 (1980)
4. Ueng, P. P., et al.: ibid. *3*, 315 (1980)
5. Chiang, L.-C., et al.: Appl. Environ. Microbiol. *42*, 284 (1981)
6. Chen, W.-P.: Proc. Biochem. *15*, 30 (1980)
7. Chen, W.-P.: ibid. *15*, 36 (1980)
8. Gong, C.-S., et al.: Appl. Environ. Microbiol. *41*, 430 (1981)
9. Chiang, L.-C., et al.: Biotech. Bioeng. Symp. *11*, 263 (1981)
10. Jeffries, T. W.: ibid *11*, 315 (1981)
11. Gong, C.-S., et al.: Adv. Biochem. Eng. *20*, 93 (1981)
12. Hsiao, H.-Y., et al.: Enzyme Microb. Technol. *4*, 25 (1982)
13. Schneider, H., et al.: Biotechnol. Lett. *3*, 89 (1981)
14. Maleszka, R., Veliky, I. A., Schneider, H.: ibid. *3*, 415 (1981)
15. Slininger, P. J., et al.: Biotech. Bioeng. *24*, 371 (1982)
16. Jeffries, T. W.: Biotechnol. Lett. *3*, 213 (1981)
17. Gong, C.-S., McCracken, L. D., Tsao, G. T.: ibid. *3*, 245 (1981)
18. Gong, C.-S., Ladisch, M. R., Tsao, G. T.: ibid *3*, 657 (1981)
19. White, M. G., Willaman, J. J.: Biochem. J. *22*, 583 (1928)
20. Sciarini, L. J., Wirth, J. C.: Cereal Chem. *22*, 11 (1944)
21. Nord, F. F., Mull, R. P.: Adv. Enzymol. Rel. Subj. Biochem. *5*, 165 (1945)

22. Gibbs, M., et al.: Arch. Biochem. *50*, 237 (1954)
23. Suihko, M. L., Enari, T.-m.: Biotechnol. Lett. *3*, 723 (1981)
24. Langlykke, A. F., Van Lanen, J. M., Fraser, D. R.: Ind. Eng. Chem. *40*, 1716 (1948)
25. Horecker, B. L.: Pentose Metabolism in Bacteria, p. 100, New York: John Wiley and Sons 1963
26. Spivey, M. J.: Proc. Biochem. *13*, 2 (1978)
27. Rosenberg, S. L.: Enzyme Microb. Technol. *2*, 185 (1980)
28. Flickinger, M. C.: Biotech. Bioeng. *22* (Supp. 1), 27 (1980)
29. Zeikus, J. G.: Ann. Rev. Microbiol. *34*, 423 (1980)
30. Rydholm, S. A.: Pulping Processes, p. 95, New York: Interscience Publishers, John Wiley and Sons 1965
31. Browning, B. L.: Composition and chemical reactions of wood, in: The Chemistry of Wood. (Browning, B. L. ed.), p. 70, New York: Interscience Publishers, John Wiley and Sons 1963
32. Sjöström, E.: Wood Chemistry, Fundamentals and Applications, p. 208, New York: Academic Press 1981
33. Sloneker, J. H.: Biotech. Bioeng. Symp. *6*, 235 (1976)
34. Krull, L. H., Inglett, G. E.: J. Agric. Food Chem. *28*, 917 (1980)
35. Wilkie, K. C. B.: Adv. Carbohyd. Chem. Biochem. *36*, 215 (1979)
36. Detroy, R. W., Hesseltine, C. W.: Proc. Biochem. *13*, 2 (1978)
37. Ashare, E., Burid, M. G., Wilson, E. H.: Feasibility Study for Anaerobic Digestion of Crop Residues. SERI/TR-8157-1. Springfield, Va.: National Technical Information Service 1979
38. USDA Forest Service: Misc. Pub. 1394. Washington, D. C.: Sup. Doc. U.S. Gov. Print Off. 1980
39. Arthur D. Little, Inc.: Use of Wood Residues. U.S. DOE/EC-77-03-1692. Cambridge, Mass.: Arthur D. Little, Inc. 1979
40. Wahlgren, H. G., Ellis, T. H.: Tappi *61*, 37 (1978)
41. Salo, D. J., Henry, J. F.: Potential Availability of Wood as a Feedstock for Methanol Production. DOE/ET-0114/1. Springfield, Va.: National Technical Information Service 1979
42. American Paper Institute, Inc.: U.S. Pulp, Paper, and Paperboard Industry Estimate Fuel and Energy Use, New York: API 1980
43. Mehlbert, R.: Hemicellulose hydrolysis and leaching, in: LORRE Biomass Conversion Conf., p. 9, West Lafayette, Ind.: Purdue Univ. 1981
44. Ackerson, M., Ziobro, M., Gaddy, J. L.: Biotech. Bioeng. Symp. *11*, 103 (1981)
45. Knappert, D., Grethlein, H., Converse, A.: ibid. *11*, 67 (1981)
46. Hajny, G. J.: Biological Utilization of Wood for Production of Chemicals and Foodstuffs. USDA For. Serv. Res. Pap. FPL 385. Madison, Wis.: Forest Products Lab. 1980
47. Azhar, A. F., et al.: Biotech. Bioeng. Symp. *11*, 293 (1981)
48. Wood, W. A.: Ann. Rev. Biochem. *35*, 521 (1966)
49. Mortlock, R. P.: Adv. Microb. Physiol. *13*, 1 (1976)
50. Sols, A.: Regulation of carbohydrate transport and metabolism in yeast, in: Aspects of Yeast Metabolism. (Mills, A. K., Krebs, H. eds), p. 45, Philadelphia: F. A. Davis Co. 1967
51. Fiechter, A., Fuhrmann, G. F., Käppeli, O.: Adv. Microb. Physiol. *22*, 123 (1981)
52. Canh, D. S., et al.: Folia Microbiol. *20*, 320 (1975)
53. Lam, V. M. S., et al.: J. Bacteriol *143*, 396 (1980)
54. Mitchell, P.: Membranes of cells and organelles: morphology, transport and metabolism, in: Organization and Control in Prokaryotic and Eukaryotic Cells. (Charles, H. P., Knight, B. C. J. G. eds.), p. 121, London: Cambridge University Press 1970
55. Lehmer, A., Schleifer, K. H.: Zbl. Bact. Hyg., I. Abt. Orig. Cl, 109 (1980)
56. London, J., Chace, N. M.: Proc. Natl. Acad. Sci. U.S.A. *74*, 4296 (1977)
57. London, J., Chace, N. M.: J. Bacteriol. *140*, 949 (1979)
58. Curtis, S. J.: ibid. *120*, 295 (1974)
59. David, J., Wiesmeyer, H.: Biochim. Biophys. Acta *208*, 45 (1970)
60. Shamanna, D. K., Sanderson, K. E.: J. Bacteriol. *139*, 64 (1979)
61. Kleinzeller, A., Kotyk, A.: Transport of monosaccharides in yeast cells and its relationship to cell metabolism, in: Aspects of Yeast Metabolism. (Mills, A. K., Krebs, H. eds.), p. 33, Philadelphia: F. A. Davis Co. 1967

62. Alcorn, M. E., Griffin, C. C.: Biochim. Biophys. Acta *510*, 361 (1978)
63. Janda, S.: Folia Microbiol. *22*, 433 (1977)
64. Janda, S., Kotyk, A., Tauchova, R.: Arch. Microbiol. *111*, 151 (1976)
65. Höfer, M., Misra, P. C.: Biochem. J. *172*, 15 (1978)
66. Heller, K. B., Hoefer, M.: Biochim. Biophys. Acta *514*, 172 (1978)
67. Hauer, R., Hoefer, M.: J. Membr. Biol. *43*, 335 (1978)
68. Janda, S.: Cell. Mol. Biol. *25*, 131 (1979)
69. Niemietz, C., Hauer, R., Höfer, M.: Biochem. J. *194*, 433 (1981)
70. Srivastava, V., Misra, P. C.: Toxicol. Lett. *7*, 475 (1981)
71. Kotyk, A., Michaljanikova, D.: Folia Microbiol. *23*, 18 (1978)
72. Miersch, J.: ibid. *22* 363 (1977)
73. Bilai, V. I., Stryzhevs'ka, A. Y.: Mikrobiol. Zh. (Kiev) *39*, 307 (1977)
74. Chiang, C., Knight, S. G.: Nature *188*, 79 (1960)
75. Tomoyeda, M., Horitsu, H.: Agric. Biol. Chem. *28*, 139 (1964)
76. Höfer, M., Betz, A., Kotyk, A.: Biochim. Biophys. Acta *252*, 1 (1971)
77. Barnett, J. A.: Adv. Carbohydr. Chem. Biochem. *32*, 125 (1976)
78. Lampen, J. O., Mitsuhashi, S.: J. Biol. Chem. *204*, 1011 (1953)
79. Slein, M. W.: J. Am. Chem. Soc. *77*, 1663 (1955)
80. Hochster, R. M., Watson, R. W.: Arch. Biochem. Biophys. *48*, 120 (1957)
81. Yomanaka, K.: Biochim. Biophys. Acta *151*, 670 (1968)
82. Vaheri, M., Kauppinen, V.: Proc. Biochem. *12*, 5 (1977)
83. Kluepfel, D., Biron, L., Ishaque, M.: Biotechnol. Lett *2*, 309 (1980)
84. Park, Y. H., Chung, T. W., Han, M. H.: Enzyme Microb. Technol. *2*, 227 (1980)
85. Wilson, B. L., Mortlock, R. P.: J. Bacteriol. *113*, 1404 (1973)
86. Horitsu, H., Sasaki, I., Tomoyeda, M.: Agric. Biol. Chem. *34*, 676 (1970)
87. Scher, B. M., Horecker, B. L.: Polyol dehydrogenases of *Candida utilis* II. TPN-linked dehydrogenase, in: Methods in Enzymology, Vol. 9, (Wood, W. A. ed.), p. 166, New York: Academic Press 1966
88. Moret, V., Sperti, S.: Arch. Biochem. Biophys. *98*, 124 (1962)
89. Scher, B. M., Horecker, B. L.: ibid. *116*, 117 (1966)
90. Horitsu, H., Tomoyeda, M., Kumagai, K.: Agric. Biol. Chem. *32*, 514 (1968)
91. Wang, S.-Y. C., LeTorneau, P.: Arch. Mikrobiol. *93*, 87 (1973)
92. Suzuki, T., Onishi, H.: Agric. Biol. Chem. *39*, 2389 (1975)
93. Karassevitch, Y. N.: Biochimie *58*, 239 (1976)
94. Barnett, J. A.: J. Gen. Microbiol. *52*, 131 (1968)
95. Lewis, D. H., Smith, D. C.: New Phytol. *66*, 143 (1967)
96. Touster, O., Shaw, D. R. W.: Physiol. Rev. *42*, 181 (1962)
97. Chakravorty, M., et al.: J. Biol. Chem. *237*, 1014 (1962)
98. Chakravorty, M., Horecker, B. L.: Polyol dehydrogenases of *Candida utilis* I. DPN-linked dehydrogenase, in: Methods in Enzymology, Vol. 9, (Wood, W. A. ed.), p. 163, New York: Academic Press 1966
99. Birken, S., Pisano, M. A.: J. Bacteriol. *125*, 225 (1976)
100. Veiga, L. A.: J. Gen. Appl. Microbiol. *14*, 65 (1968)
101. Karasevich, Yu. N.: Microbiology *39*, 649 (1970)
102. Karasevich, Yu. N., Ipatova, A. P.: ibid. *37*, 167 (1968)
103. Watson, J. A., et al.: J. Bacteriol. *100*, 110 (1969)
104. Suzuki, T., Onishi, H.: Appl. Microbiol. *24*, 850 (1973)
105. Yoshitake, J., et al.: Agric. Biol. Chem. *40*, 1493 (1976)
106. Mortlock, R. P., Wood, W. A.: J. Bacteriol. *88*, 838 (1964)
107. Mortlock, R. P., Wood, W. A.: ibid. *88*, 845 (1964)
108. Kersters, K., DeLey, L.: Polyol dehydrogenases of *Gluconobacter*, in: Methods in Enzymology, Vol. 9, (Wood, W. A. ed.), p. 170, New York: Academic Press 1966
109. Fossitt, D. P., Wood, W. A.: Pentitol dehydrogenases of *Aerobacter aerogenes*, in: Methods in Enzymology, Vol. 9, (Wood, W. A. ed.), p. 180, New York: Academic Press 1966
110. Andrejew, A.: C. R. Acad. Sci. Paris, *289*: Serie D: 1241–1244 (1979)
111. Yamanaka, K., Gino, M., Kaneda, R.: Agric. Biol. Chem. *41*, 1493 (1977)
112. Yamanaka, K., Gino, M.: Hakkokogaku Kaishi *57*, 322 (1979)
113. Gong, C.-S., Chen, L. F., Tsao, G. T.: Biotechnol. Lett. *3*, 125 (1981)

114. Mitsuhashi, S., Lampen, J. O.: J. Biol. Chem. *204*, 1011 (1953)
115. Stumpf, P. K., Horecker, B. L.: ibid. *218*, 753 (1956)
116. Neuberger, M. S., Hartley, B. S., Walker, J. E.: Biochem. J. *193*, 513 (1981)
117. Wang, P. Y., Schneider, H.: Can. J. Microbiol. *26*, 1165 (1980)
118. Cochrone, V. W.: Glycolysis, in: The Filamentous Fungi, Vol. 2, (Smith, J. E., Berry, D. R. eds.), p. 70, New York: John Wiley 1976
119. Kiely, M. E., Tan, E. L., Wood, T.: Can. J. Biochem. *47*, 455 (1969)
120. Clark, M. G., Williams, J. F., Blackmore, P. F.: Biochem. J. *125*, 381 (1971)
121. Cavalieri, S. W., Neet, K. E., Sable, H. Z.: Arch. Biochem. Biophys. *171*, 527 (1975)
122. Klein, H., Brand, K.: Hoppe-Seyler's Z. Physiol. Chem. *358*, 1325 (1977)
123. Neto, J. S. A., Panek, A. D.: Arch. Biochem. Biophys. *194*, 354 (1979)
124. Davis, L., Lee, N., Glaser, L.: J. Biol. Chem. *247*, 5862 (1972).
125. Quayle, J. R., Ferenci, T.: Microbiol. Rev. *42*, 251 (1978)
126. Williams, J. F., Blackmore, P. F., Clark, M. G.: Biochem. J. *176*, 257 (1978)
127. Williams, J. F.: TIBS *5*, 315 (1980)
128. Katzl, J.: TIBS *6*, xiv (1981)
129. O'Connor, M. L., Quayle, J. R.: J. Gen. Microbiol. *120*, 219 (1980)
130. Horecker, B. L. et al.: Comparative studies of aldolases and fructose diphosphatases, in: Aspects of Yeast Metabolism. (Mills, A. K., Krebs, H. eds.), p. 71, London: F. A. Davis Co. 1967
131. Höfer, M., et al.: Biochem. J. *123*, 855 (1971)
132. Whitworth, D., Ratledge, C.: J. Gen. Microbiol. *102*, 397 (1977)
133. Sgorbati, B., Lenaz, G., Casalicchio, F.: Antonie van Leeuwenhoek *42*, 49 (1976)
134. Thauer, R. K., Jungermann, K., Decker, K.: Bacteriol. Rev. *41*, 100 (1977)
135. Heath, E. C., et al.: J. Biol. Chem. *231*, 1009 (1958)
136. Fred, E. B., Peterson, W. H., Anderson, J. A.: ibid. *48*, 385 (1921)
137. Lampen, J. O., Gest, H., Sowden, J. C.: J. Bacteriol. *61*, 97 (1951)
138. Rappaport, D. A., Barker, H. A., Hassid, W. Z.: Arch. Biochem. Biophys. *31*, 326 (1951)
139. Bernstein, I. A.: J. Biol. Chem. *205*, 309 (1953)
140. Gibbs, M., et al.: Archiv. Biochem. *50*, 237 (1954)
141. De Moss, R. D., Bard, R. C., Gunsalus, I. C.: J. Bacteriol. *62*, 449 (1951)
142. De Moss, R. D., Gunsalus, I. C., Bard, R. C.: ibid. *66*, 10 (1953)
143. Dobrogosz, W. J., De Moss, R. D.: ibid. *85*, 1356 (1963)
144. Dobrogosz, W. J., De Moss, R. D.: Biochim. Biophys. Acta *77*, 629 (1963)
145. Lee, C. K., Dobrogosz, W. J.: J. Bacteriol. *90*, 653 (1965)
146. Devries, W., Stouthamer, A. H.: ibid. *93*, 574 (1967)
147. Turner, K. W., Roberton, A. M.: Appl. Environ. Microbiol. *38*, 7 (1979)
148. Greenley, D. E., Smith, D. W.: Arch. Microbiol. *122*, 257 (1979)
149. Wood, A. P., Kelly, D. P.: J. Gen. Microbiol. *120*, 333 (1980)
150. Ratledge, C., Botham, P. A.: ibid. *102*, 391 (1977)
151. Botham, P. A., Ratledge, C.: ibid. *114*, 361 (1979)
152. Goldberg, M. L., Racker, E.: J. Biol. Chem. *237*, 3841 (1962)
153. Schröter, W., Holzer, H.: Biochim. Biophys. Acta *77*, 474 (1963)
154. Votaw, R. G., Krampitz, L. O.: Fed. Proc. Fed. Amer. Soc. Exp. Biol. *25*, 342 (1966)
155. Cristen, P., Gasser, A.: Eur. J. Biochem. *107*, 73 (1980)
156. Yu, E. K. C., Saddler, J. N.: Biotechnol. Lett. *4*, 121 (1982)
157. Phaff, H. J., Miller, M. W., Mrak, E. M.: The Life of Yeasts, p. 136, Cambridge, Mass.: Harvard University Press 1978
158. Lagunas, R.: TIBS *6*, 201 (1981)
159. Fiechter, A.: Proceeding of the Fourth Int. Symp. on Yeasts, Part II, p. 17, Vienna, Austria 1974
160. Scheffers, W. A.: Nature *210*, 533 (1966)
161. Wikén, T. O.: On "negative Pasteur effects" in yeasts, in: Aspects of Yeast Metabolism (Mills, A. K., Krebs, H. eds.), p. 133, Philadelphia: F. A. Davis Co. 1967
162. Wikén, T. O., Scheffers, W. A., Verhaar, A. J. M.: Antonie van Leeuwenhoek *24*, 401 (1961)
163. Scheffers, W. A., Wikén, T. O.: ibid. *35A*, 31 (1969)
164. Scheffers, W. A.: Experientia *17*, 40 (1961)
165. Carrascosg, J. M., et al.: Antonie van Leeuwenhoek *47*, 209 (1981)

166. Sims, A. P., Barnett, J. A.: J. Gen. Microbiol. *106*, 277 (1978)
167. Jeffries, T. W.: Biotech. Bioeng. Symp. *12*, 103 (1983)
168. Turner, J. F., Turner, D. H.: Ann. Rev. Plant Physiol. *26*, 159 (1975)
169. Bonsignore, A., DeFlora, A.: Curr. Topics Cell. Regula. *6*, 21 (1972)
170. Yue, R. H., Noltman, E. A., Kuby, S. A.: J. Biochem. *244*, 1353 (1969)
171. Eggleston, L. V., Krebs, H.: Biochem. J. *138*, 425 (1974)
172. Kuby, S. A., Wu, J. T., Roy, R. N.: Arch. Biochem. Biophys. *165*, 153 (1974)
173. Passonneau, J. V., Schulz, D. W., Lowry, O. H.: Fed. Proc. *25*, 219 (1966)
174. Levy, R. H.: Adv. Enzymol. *48*, 97 (1979)
175. Polakis, E. S., Bartley, W.: Biochem. J. *99*, 521 (1966)
176. Grove, T. H., Ishaque, A., Levy, H. R.: Arch. Biochem. Biophys. *177*, 307 (1976)
177. Olive, C., Geroch, M. E., Levy, H. R.: J. Biol. Chem. *246*, 2047 (1971)
178. Benziman, M., Mazover, A.: ibid. *248*, 1603 (1973)
179. Avigad, G.: Proc. Nat. Acad. Sci., U.S. *56*, 1543 (1966)
180. Bonsignore, A., et al.: Ital. J. Biochem. *15*, 458 (1966)
181. Domagk, G. F., Chilla, R., Doering, K. M.: Life Sci. *13*, 655 (1973)
182. Holzer, H., Witt, I.: Biochim. Biophys. Acta *38*, 163 (1960)
183. Onishi, H., Saito, N., Koshiyama, I.: Agric. Biol. Chem. *25*, 124 (1961)
184. Onishi, H., Suzuki, T., Ouchi, T.: ibid. *44*, 35 (1980)
185. Osmond, C. B., Rees, T. A.: Biochim. Biophys. Acta *184*, 35 (1969)
186. Handinson, O., Cove, D. J.: J. Biol. Chem. *249*, 2344 (1974)
187. Jessup, W., Fouler, M. W.: Planta *137*, 71 (1977)
188. Gong, C.-S., et al.: Biotech. Bioeng. *25*, 85 (1983)
189. Malezka, R., Schneider, H.: Can. J. Microbiol. *28*, 360 (1982)
190. Anderson, A. K., Willaman, J. J.: Proc. Soc. Exptl. Biol. Med. *20*, 108 (1922)
191. Chiang, L.-C., et al.: Enzyme Microb. Technol. *4*, 93 (1982)
192. Chiang, L.-C., et at.: Appl. Environ. Microbiol. *42*, 66 (1981)
193. Van Uden, N., Vidal-Leivia, M.: *Torulopsis berlese*, in: The Yeasts. (Lodder, J. ed.), p. 1249, Amsterdam–London: North Holland 1971[2]
194. Veng, P. P., Gong, C.-S.: Enzyme Microb. Technol. *4*, 169 (1982)
195. Hsiao, H.-y., et al.: Appl. Environ. Microbiol. *43*, 840 (1982)
196. Maleszka, R., et al.: Biotechnol. Lett. *4*, 133 (1982)
197. Swings, J., Deley, J.: Bacteriol. Rev. *41*, 1 (1977)
198. Dekker, R. F. H., Richards, G. N.: Adv. Carbohyd. Chem. Biochem. *32*, 277 (1976)
199. Biely, P., et al.: Folia Microbiol. *23*, 366 (1978)
200. Avgerinos, G. C. et al.: A novel single-step microbial conversion of cellulosic biomass to ethanol, in: Advances in Biotechnology Vol. II, (Moo-Young, M., Robinson, C. W. eds.), p. 119, Toronto: Pergamon Press 1981
201. Alexander, J. K., Connors, R., Yamamoto, N.: Production of liquid fuels from cellulose by combined saccharification-fermentation or cocultivation of clostridia, in: Advances in Biotechnology Vol. II, (Moo-Young, M., Robinson, C. W. eds.), p. 119, Toronto: Pergamon Press 1981
202. Ng, T. K., Ben-Bassat, A., Zeikus, J. G.: Appl. Environ. Microbiol. *41*, 1337 (1981)
203. Brooks, R., et al.: Bioconversions of plant biomass to ethanol, in: 3rd Ann. Biomass Energy Conf. Proceedings. SERI/TP-33-285. p. 275. Springfield, Va.: National Technical Information Service 1979
204. Andreesen, J. R., et al.: J. Bacteriol. *114*, 743 (1973)
205. Schwartz, R. D., Keller, F. A., Jr.: Appl. Environ. Microbiol. *43*, 117 (1982)
206. Targonski, Z., Szajer, C.: Biotechnol. Lett. *1*, 75 (1979)
207. Targonski, Z., Szajer, C.: ibid. *1*, 439 (1979)
208. Trivedi, L. S., Rao, K. K.: ibid. *3*, 481 (1981)
209. Veng, P. P., Gong, C.-S.: Plant Physiol. *65*, 6 (1980)
210. Gong, C.-S., Mann, C. M., Tsao, G. T.: Biotechnol. Lett. *3*, 77 (1981)
211. Avgerinos, G. C., Wang, D. I. C.: Annu. Reports Ferm. Proc. *4*, 165 (1980)

D-Xylose Metabolism by Mutant Strains of *Candida* sp.

Linda D. McCracken[1] and Cheng-Shung Gong[2]
Laboratory of Renewable Resources Engineering, Purdue University, West Lafayette, Indiana 47907

The first step in the metabolism of D-xylose by yeasts and mycelial fungi was found to be the reduction of D-xylose to xylitol, a reaction catalyzed by NADPH-linked D-xylose reductase. This step is followed by the oxidation of xylitol to D-xylulose which is catalyzed by NAD-linked xylitol dehydrogenase. This oxidation-reduction reaction appears to be an obligatory step in D-xylose metabolism since direct isomerization of D-xylose to D-xylulose does not occur. The D-xylulose formed is then phosphorylated to D-xylulose-5-phosphate, the key intermediate. Similar metabolic routes have also been suggested for other aldopentoses. This enzyme system in yeasts enables them to assimilate pentoses to produce pentitols as the major metabolic products. Recently yeast mutants that exhibit different product patterns have been isolated. Studies of the enzyme activities of a xylose-utilizing yeast, *Candida* sp. C2, and its ethanol-producing mutant XF217 have been conducted. The specific activities of xylitol dehydrogenase and xylulokinase in XF217 increased significantly over those of the parent strain, C2. D-Xylose reductase activity remained the same. The increased xylitol dehydrogenase and xylulokinase activities enable this strain to shift from xylitol to ethanol production. Instead of excreting xylitol as a final product, more xylitol is converted to D-xylulose and ultimately to ethanol.

[1] Present address: Department of Plant Pathology, Pennsylvania State University, University Park, PA 16802;
[2] To whom all correspondence should be addressed

1 Introduction

The pentose sugars, D-xylose and L-arabinose, comprise up to 30% of the neutral carbohydrates derived from woods, agricultural crop residues, and other plant materials. These sugars along with D-glucose and other minor hexoses reside in secondary cell wall materials that are often referred to as "hemicellulose".

Unlike the orderly crystalline structure of cellulose, hemicellulose exhibits variability in both structure and sugar constitutents. The degree of polymerization of hemicellulose is usually less than 200 [1,2] whereas native cellulose has a degree of polymerization greater than 1,000 [3].

In nature the hemicellulose portion of the plant cell wall is often degraded by microbial action faster than the cellulose portion. Schmidt and Walter [4] studied the succession and activity of microorganisms in metabolism of stored sugarcane bagasse; first the residual soluble sugars are consumed, followed by pectic materials, hemicellulose, cellulose and finally lignins. Yeasts, mycelial fungi and bacteria are all associated with the hemicellulose degradation. Many yeast species are able to synthesize extracellular hemicellulases that are responsible for hemicellulose degradation [5]. As a result it is not surprising that many yeast species can utilize D-xylose and L-arabinose, hemicellulose degradation products, quite readily. Barnett [6] surveyed a total of 434 strains of yeast and found that about half of those surveyed (214 strains) could utilize and metabolize D-xylose and L-arabinose aerobically. Utilization of these two sugars often results in the production of xylitol and L-arabitol as the main metabolic products [7,8]. This is due to enzyme systems in yeasts (e.g., pentose reductase) that enable them to convert pentoses to corresponding pentitols [9].

In addition to pentitols production from pentoses, many yeasts are capable of producing polyols from hexoses. The most common polyols derived from hexoses are D-mannitol, sorbitol, dulcitol, erythritol and glycerol. Polyols in yeasts serve the following functions:

a) as storage carbohydrates;
b) storage of reducing power;
c) coenzyme regulation;
d) osmoregulation;
e) enzyme protection under low water activity environments.

For detailed discussion of these subjects many review articles are available [10-13].

Many useful products, including ethanol, can be produced from hexoses and pentoses by microorganisms. Yeasts, in general, are the better choice for conversion of carbohydrates to ethanol than are other microorganisms due to their high rate, ethanol tolerance, and high ethanol yields. In the design of a simple industrial process to produce ethanol from biomass-derived carbohydrates, a biological system that could degrade both pentoses and hexoses simultaneously to produce high yields of ethanol with high rates of conversion is essential.

Ethanol was not recognized as a metabolic product from D-xylose by yeasts until recently. Gong et al. [14] reported that a small quantity of ethanol is produced from D-xylose by several yeast species even though xylitol is the major product. Further examination of many yeast strains by Gong et al. revealed that a wide range of yeasts including those belonging to the genera *Saccharomyces* and *Schizosaccharomyces* are able to produce ethanol from D-xylose [8]. Other yeasts that have been studied with respect to ethanol production from D-xylose include a strain of *Candida tropicalis* [15], *Pachysolen tannophilus* [16,17], and a mutant yeast, *Candida* sp., [18]. The detailed discussion of ethanol production from D-xylose by yeasts has recently been reviewed [19,20].

In this article, we review the enzymes involved in pentose metabolism by yeasts and also described the recent results with respect to D-xylose metabolism in a *Candida* yeast and its mutants.

2 Pentose Metabolism

Enzymatic studies of cell-free extracts of a mycelial fungi, *Penicillium chrysogenum*, by Chiang and Knight has shown that aldopentose is reduced to pentitol by a NADPH-linked reductase, this reaction is followed by the oxidation of the pentitol to either D-xylulose or L-xylulose by a NAD-linked dehydrogenase [21-24]. They proposed the following pathway for D-xylose and L-arabinose metabolism by *P. chrysogenum*:

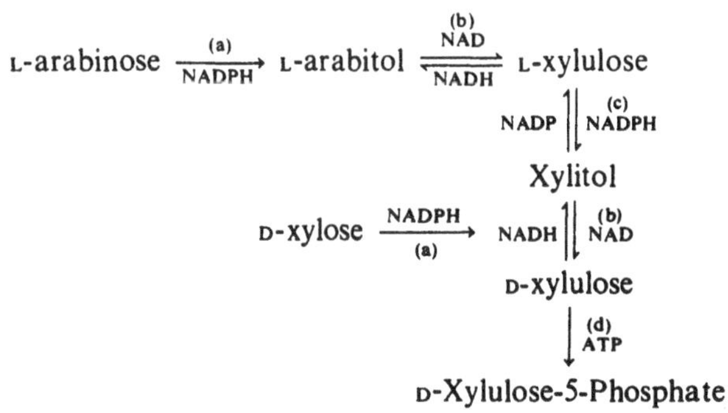

The enzymes involved are aldopentose reductase (a), pentitol-NAD dehydrogenase (b), L-xylulose-NADPH reductase (c), and D-xylulokinase (d). They also [25] surveyed 20 strains of yeasts and mycelial fungi for their ability to grow in D-xylose and found that the 14 strains that were capable of growth all possessed the enzymatic systems to convert D-xylose to D-xylulose through xylitol (Table 1).

Veiga et al. [26] studied pentose metabolism in the yeast, *Candida albicans*, and found that a NADP-linked dehydrogenase is responsible for the interconversion of aldopentose to its corresponding pentitol. No xylose isomerase has been detected in cell-free extracts of this yeast. They concluded that ketopentose is formed from aldopentose by way of the sugar alcohol rather than by direct isomerization. This assumption was tested and confirmed by Chakravorty et al. [27] in their studies of *C. utilis*, Moret and Sperti [28] in their studies of *Geotrichum candidum*, and Suzuki and Onishi [29] in their analysis of *Pichia quercuum*.

Nevertheless, the presence of xylose isomerase in *C. utilis* [30] and an obligatory aerobic yeast, *Rhodotorula gracilis* has been reported [31]. Tomoyeda and Horitsu [30] purified a D-xylose isomerase from D-xylose grown *C. utilis* and found this enzyme

Table 1. Specific activities of xylose reductase and xylitol dehydrogenase in mycelial fungi and yeasts [25]

Organisms	Specific activities[a]	
	Xylose reductase	Xylitol dehydrogenase
Mycelial Fungi		
Aspergillus fumigatus	0.32	0.084
Aspergillus ochraceus	0.23	0.065
Aspergillus oryzae	0.33	1.41
Aspergillus niger	0.65	0.1
Byssochlamys fulva	1.01	0.081
Gliocladium roseum	1.71	0.198
Myrothecium verrucaria	2.89	0.256
Neurospora crassa	0.40	0.079
Penicillium chrysogenum	2.32	0.217
Penicillium citrinum	2.33	0.261
Penicillium expansum	2.34	0.136
Yeasts		
Rhodotorula glutinis	1.02	0.071
Torulopsis utilis (*Candida utilis*)	1.39	0.112

[a] Micromoles substrates reduced or oxidized per mg protein

shared many characteristics similar with that prepared from bacteria. Höfer et al. [31] observed D-xylose isomerase activity in D-xylose grown cell-free extracts of *R. gracilis* and concluded that the direct isomerization of D-xylose to D-xylulose catalyzed by D-xylose isomerase is an obligatory step in this yeast for D-xylose metabolism. In contrast, Chakravorty et al. [27] failed to detect any D-xylose isomerase activity in D-xylose grown cell-free extracts of *C. utilis*. Similarly Veiga [32] was unable to detect any isomerase activity in *C. albicans*.

A variety of metabolic products can be produced by yeasts from pentoses. These include several polyols, ethanol, and organic acids. The pathways leading to products are summarized in Fig. 1.

2.1 D-Xylose Reductase

Xylose reductase is generally known as NADP-linked polyol dehydrogenase, which is distinguishable from another group of enzymes, NAD-linked polyol dehydrogenase. Chiang and Knight [22] described a polyol dehydrogenase from cell-free extract of *Penicillium chrysogenum*. They called it xylose reductase because of its relative substrate specificity toward D-xylose, although good activity was also observed with L-arabinose. D-Ribose and six carbon sugars such as D-glucose and D-galactose can also serve as substrates.

This enzyme is capable of catalyzing the reduction of aldoses to corresponding polyols and requires reduced NADP to carry out electron transfer. Numerous investigators have reported enzymes of similar nature from strains of mycelial fungi and yeasts which are capable of catalyzing the reduction of D-xylose to

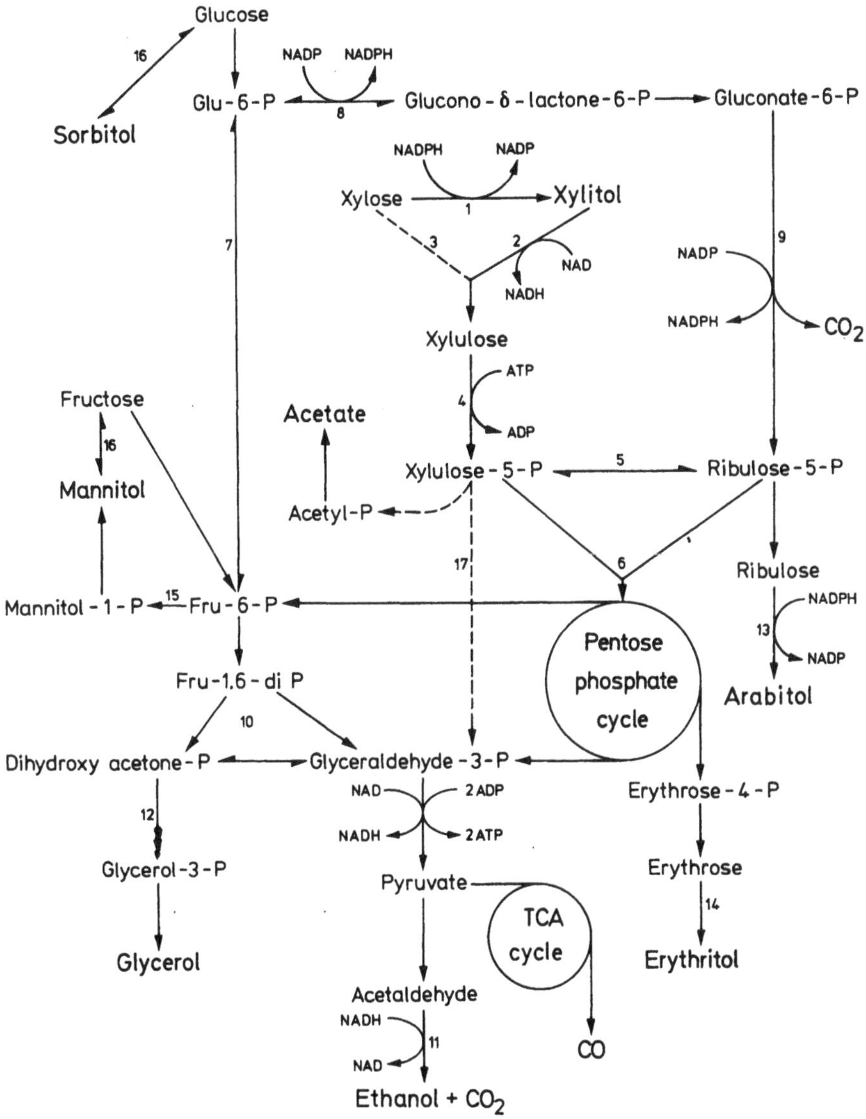

Fig. 1. D-Xylose and D-glucose metabolism by yeasts. 1, xylose reductase (aldoreductase); 2, xylitol dehydrogenase (xylulose reductase); 3, xylose isomerase; 4, xylulokinase; 5, phosphoketopentoepimerase; 6, transaldolase and transketolase; 7, phosphohexoisomerase; 3, glucose-6-phosphate dehydrogenase; 9, phosphogluconate dehydrogenase; 10, aldolase; 11, alcohol dehydrogenase; 12, glycerol-3-phosphate dehydrogenase; 13, arabitol dehydrogenase; 14, erythritol dehydrogenase; 15, mannitol-1-phosphate oxidoreductase; 16, polyol dehydrogenase; 17, xylulose-5-phosphate phosphoketolase

xylitol. Such as enzyme has been described for *C. albicans* [26,32], *C. utilis* [33], *Geotrichum candidum* [28], *Pichia quercuum* [29], *Cephalosporium chrysogenum* [34], *Melampsora lini* [35] and for those yeasts studied by Chiang and Knight [25] (Table 1).

The existence of this enzyme in eukaryotic microorganisms is probably common in view of the fact that many yeasts studied by Onishi and Suzuki [7] and Gong et al. [8] are capable of producing xylitol from D-xylose.

Very few xylose reductases have been purified and studied in detail. Veiga [32] purified this enzyme (134-fold) from *C. albicans* and found that a wide variety of pentoses and hexoses can be used as substrates (Table 2). Based on the substrate specificity Veiga concluded that aldose in the D-glycero-configuration with a hydroxyl group attached to carbon-2 are the substrates of xylose reductase. Sugars lacking hydroxyl group at carbon-2 are poor substrates. The reverse reaction using polyols

Table 2. Substrate specificity of NADPH-polyol dehydrogenase from yeasts and mycelial fungi

Substrates	Relative activity (%)[a]			
	Candida albicans [32]	*Candida utilis* [33]	*Melampsora lini* [35]	*Penicillium chrysogenum* [22]
D-Xylose	100	100	100	100
L-Xylose	5	0	—	0
D-Arabinose	13	0	—	0
L-Arabinose	76	216	118	25
D-Ribose	71	69	77	25
D-Lyxose	11	0	—	—
D-Glucose	36	27	14	—
D-Galactose	69	81	12	—
D-Mannose	5	0	10	—
D-Fructose	2	0	—	—

[a] D-Xylose = 100%

Table 3. Substrate specificity of NADP-polyol dehydrogenase

Substrate	Relative activity (%)[a]
	C. albicans [32]
Xylitol	100
D-Arabitol	14
L-Arabitol	105
Ribitol	46
Dulcitol	121
Sorbitol	73
Mannitol	16
Erythritol	33

[a] Xylitol = 100%

as substrate with NADP as co-factor was also studied by Veiga [32]. The results indicated that several polyols are good substrates (Table 3). The enzymatic reaction product from the oxidation of xylitol was isolated and identified as D-xylose [26].

2.2 Xylitol Dehydrogenase (NAD-Polyol Dehydrogenase)

The oxidation of polyhydric alcohols to ketoses during the growth of *Acetobacter xylinum* was first described by Betrand in 1898 [36]. It was not until 1951, however, an enzyme derived from rat liver which catalyzed such a reaction was reported by Blakley and named polyol dehydrogenase [37]. In 1954, McCorkindale and Edson [38] described a polyol dehydrogenase isolated from rat liver which could catalyze the oxidation of xylitol, sorbitol, and L-iditol. Enzymes having similar activities have sometimes been referred to as sorbitol dehydrogenases or polyol dehydrogenases since they were capable of catalyzing reactions leading to the production of a wide variety of polyols [39].

There have been a number of reports of the presence of enzyme activity of this nature in bacteria (see review by Mortlock, [40]).

Numerous reports have described enzymes of a similar nature from strains of mycelial fungi and yeasts that are capable of catalyzing the oxidation of xylitol to D-xylulose. Arcus and Edson [41] reported in 1956 that cell-free extract of *C. utilis* could oxidize mannitol, sorbitol, L-rhamnitol and xylitol. Chiang and Knight [22] described the initial steps of D-xylose metabolism in *P. chrysogenum* as the reduction of D-xylose to xylitol by NADPH-linked reductase, followed by NAD-linked oxidation of xylitol to D-xylulose. In their subsequent study, they found that the mycelial fungi and yeasts that were capable of utilizing D-xylose as a carbon and energy source possessed enzymatic activities enabling them to catalyze the reduction of xylitol to xylulose (Table 1). The existence of such an enzyme system in microorganisms is not surprising since xylitol and other polyols, such as sorbitol and mannitol, are found in significant quantities in fruits, berries, and vegetables [42,43]. Xylitol dehydrogenase has been described in the cell-free extracts of *C. utilis* [27,44], *C. albicans* [45], *Pullularia pullulans* [46] and *Cephalosporium chrysogenum* [34].

Table 4. Substrate specificity of NAD-polyol dehydrogenase

Substrates	Relative activity (%)[a]			
	Candida albicans [45]	*Candida utilis* [44]	*Candida utilis* [27]	*Pullularia pullulans* [46]
Xylitol	100	100	100	100
L-Arabitol	5	9.2	1	0
D-Arabitol	2	—	7	0
D-Ribitol	50	0	23	0
D-Mannitol	4	7.4	23	0
Sorbitol	72	44.4	87	0
Erythritol	—	—	1	0

[a] Xylitol = 100%

Table 5. Substrate specificity of NADH-polyol dehydrogenase

Substrates	Relative activity (%)[a]	
	C. albicans[45)]	*C. utilis*[27)]
D-Xylulose	100	100
L-Xylulose	—	13
D-Ribulose	30	37
L-Ribulose	10	0
D-Fructose	21	53
L-Sorbose	5	0
D-Sorbose	—	0

[a] D-Xylulose = 100%

Chakravorty et al.[27)] purified xylitol dehydrogenase (30-fold) from *C. utilis* and showed this enzyme exhibited no activity toward xylitol when NADP, NADPH, or NADH were present as co-factors. It has a pH optima of 9.0 and exists as a single enzyme for the oxidation of xylitol. From the substrate specificity study, it appears that xylitol and D-xylulose are the natural substrates for this enzyme. At neutral pH, the equilibrium favors the formation of polyols.

Veiga[45)] purified xylitol dehydrogenase (20-fold) from *C. albicans* and reported that this enzyme catalyzed the oxidation of xylitol to D-xylulose, ribitol to D-ribulose and sorbitol to D-fructose. The substrate specificity of xylitol dehydrogenase from yeasts and mycelial fungi is listed in Table 4. The substrate specificity for the reverse reaction with NADH as the co-factor for polyol dehydrogenase from *C. albicans* and *C. utilis* is listed in Table 5.

Birken and Pisano[34)] purified a polyol dehydrogenase (178 fold) from a mycelial fungi *Cephalosporium chrysogenum* and found NAD-linked activity when either xylitol or sorbitol was used as a substrate. No such activity was found in response to mannitol. On the other hand, NADP-linked activity is associated with mannitol and to a lesser extent with sorbitol and D-arabitol but not with xylitol. This indicates that both NAD- and NADP-linked polyol dehydrogenase activities are associated with a single enzyme and substrate specificity is dependent on co-factors.

It is noteworthy that even though xylitol is the common intermediate of D-xylose metabolism in eukaryotic microorganisms it is not a good substrate[8)]. The inability of yeasts to utilize xylitol effectively is probably due to the limited permeability of the cell membrane to xylitol.

2.3 D-Xylulokinase

Very little information is currently available about D-xylulokinase from mycelial fungi and yeasts. Most studies of this enzyme have been of that purified from bacteria.

D-Xylulokinase has been purified from *Lactobacillus pentosus*[47)] and *Aerobacter aerogenes*[48)]. The enzyme from *A. aerogenes* is specific for D-xylulose. Other pentuloses, pentoses, hexoses and pentitols tested were not phosphorylated, except for xylitol at a slow rate. ATP serves as the phosphate donor, GTP and UTP can partially replace ATP.

D-Xylulokinase activity has been detected in cell-free extracts of *P. chrysogenum* [23] and *C. utilis* [27] when the cells were grown either in D-xylose or L-arabinose. The indirect evidence of the presence of D-xylulokinase in yeasts and mycelial fungi is provided by the fact that many of these microorganisms, including those that are not able to metabolize D-xylose, utilize D-xylulose readily (for review see Gong et al. [19] and Gong [20]. It is, therefore, suggested that D-xylulokinase is either present in yeast cells as a constitutive enzyme or it is very rapidly induced by the presence of appropriate substrates.

3 Conversion of D-Xylose by Yeast Mutants

Although basic studies on the genetics and molecular biology of yeasts have drastically expanded our understanding of the regulation of carbohydrate metabolism in a few well studied yeasts, little attention has been focused on the molecular and genetic aspects of pentose metabolism.

The potential practicality of utilizing pentoses as substrates for bioconversion by yeasts to useful products has only recently started to be appreciated. Such interest could lead to research that would offer a better understanding of the regulation of pentose metabolism in yeasts.

In this section, the recent results from the metabolic studies of yeast mutants derived from *Candida* yeast is described.

3.1 Isolation of *Candida* Yeast Mutants

Mutation and selection are the best methods for improving ethanol yields from D-xylose in yeasts. The most effective mutagens are N-methyl-N′-nitro-N-nitrosoguanidine (NG) and UV radiation. Selection via the use of metabolic analogues can also be employed to isolate mutants which are derepressed, resistant to feedback inhibition or altered in the regulation of branched or secondary metabolic pathways.

The isolation of yeast mutants with enhanced abilities to produce ethanol from D-xylose has been achieved in our laboratory [18]. Mutagenesis was carried out by UV irradiation. After UV treatment, yeast cells were spread on agar plates containing D-xylose as the sole carbon source. Colonies which appeared on the plates after 2 to 3 d of incubation at 30 C were replica plated to both D-xylose plates and xylitol plates. Those colonies that exhibited different growth rates on the different media were selected and purified by subcloning, and tested further. It should be noted that sensitivity toward UV radiation varies with different strains of yeasts. Several hundred colonies were isolated, and mutant strains were selected after an initial rapid screen for their ability or inability to utilize D-xylose for conversion to products.

3.2 Characteristics of Mutants

The mutants were initially characterized by their ability to utilize D-xylose aerobically for the formation of various products. The rates of D-xylose utilization and the products formed were analyzed and quantified by low-pressure liquid chromatography [49,50]. Typical liquid chromatography retention times for the compounds of interest are shown in Fig. 2. The actual retention times varied somewhat from one column to another.

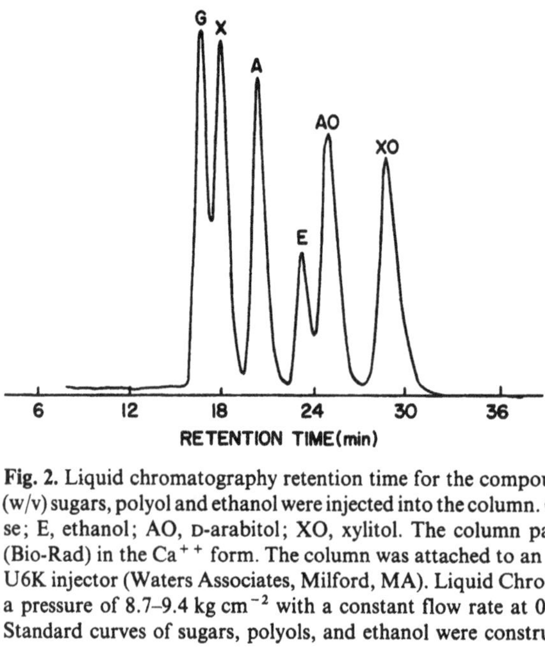

Fig. 2. Liquid chromatography retention time for the compounds of interest. Mixture of one percent (w/v) sugars, polyol and ethanol were injected into the column. G, D-glucose; X, D-xylose; A, L-arabinose; E, ethanol; AO, D-arabitol; XO, xylitol. The column packing material was Aminex (50W-X4) (Bio-Rad) in the Ca^{++} form. The column was attached to an ALC/GPC 200 series instrument with a U6K injector (Waters Associates, Milford, MA). Liquid Chromatography analysis were carried out at a pressure of 8.7–9.4 kg cm^{-2} with a constant flow rate at 0.5 ml min^{-1}, using water as the eluent. Standard curves of sugars, polyols, and ethanol were constructed

Yeast cells were grown aerobically in yeast extract-malt extractpeptone media with 1% D-xylose as the sole carbon source. After incubation which allowed for maximum growth of the yeast cells, a concentrated sterile sugar solution was added to a final concentration of 10% (w/v) D-xylose and incubated aerobically on a incubator-shaker at 200 rpm, 30 C. In another experiment, immobilized D-xylose isomerase from *Bacillus* sp. (Sweetzyme type Q, NOVO Industries, Inc.) in the amount of 30 g per liter was added to carry out isomerization-fermentation of D-xylose under oxygen-limited fermentative conditions [14].

Figure 3 depicts the liquid chromatography profile of 7 representative yeast mutants with respect to product formation from the aerobic utilization of D-xylose. The parent strain *Candida* sp. C2 is characterized by its ability to utilize D-xylose aerobically to produce xylitol as the major product with small amounts of glycerol, D-arabitol, and ethanol also being produced (Fig. 3A). The level of xylitol produced is increased in C24 (Fig. 3B) due to its enhanced rate of xylose utilization. A small increase in ethanol production was observed in one mutant, C210 (Fig. 3C).

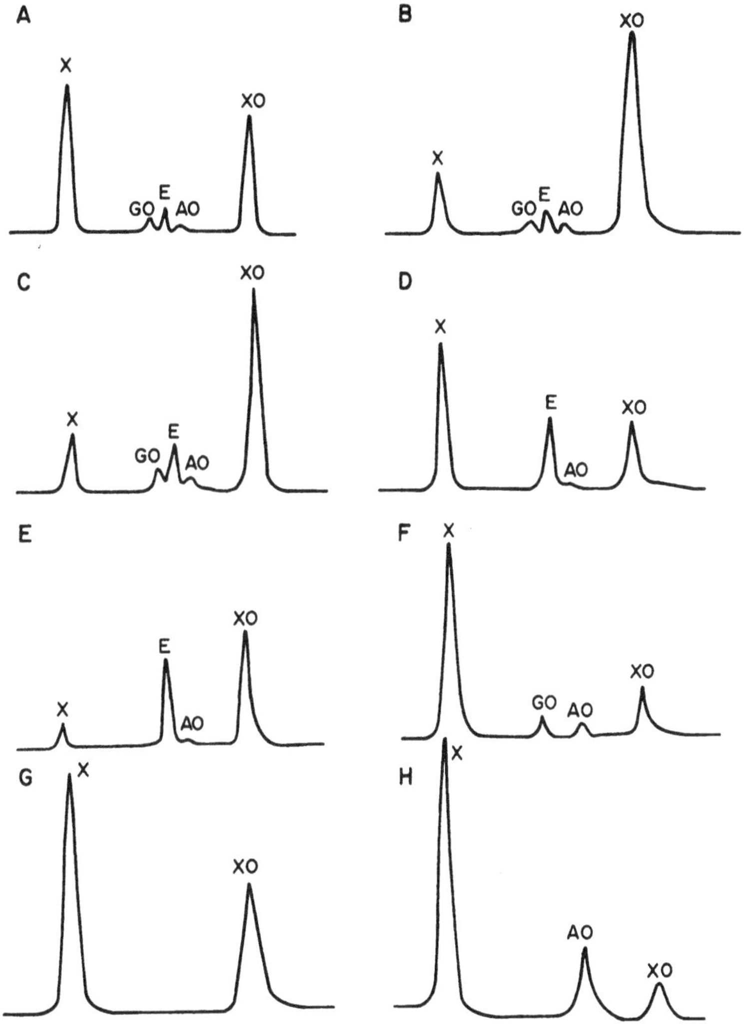

Fig. 3. Liquid chromatograms of D-xylose conversion by *Candida* sp. C2 and mutants. A, wild-type C2; B, C24; C, C210; D, C28; E, XF217; F, C232; G, C282; H, C26. X, D-xylose; GO, glycerol; E, ethanol; AO, D-arabitol; XO, xylitol. Incubation was carried out at 30 C for 36 h, and the initial D-xylose concentration was 10% (w/v)

A much greater increase in ethanol production from xylose was also observed in mutant C28 (Fig. 3D). The maximum increase in ethanol production was observed in mutant XF217 (Fig. 3E). One mutant produced glycerol and arabitol, but no ethanol (Fig. 3F). The level of arabitol was enhanced in another mutant, C26, at the expense of xylitol (Fig. 3H). The converse pattern was observed in another mutant, C282, glycerol and arabitol production were absent (Fig. 3G), but xylitol was produced.

Based on the metabolic products attained from D-xylose, these mutants fall into five basic categories e.g. those producing:

a) xylitol;
b) xylitol and ethanol;
c) xylitol, ethanol, glycerol, and arabitol;
d) xylitol and arabitol;
e) glycerol, arabitol, and xylitol.

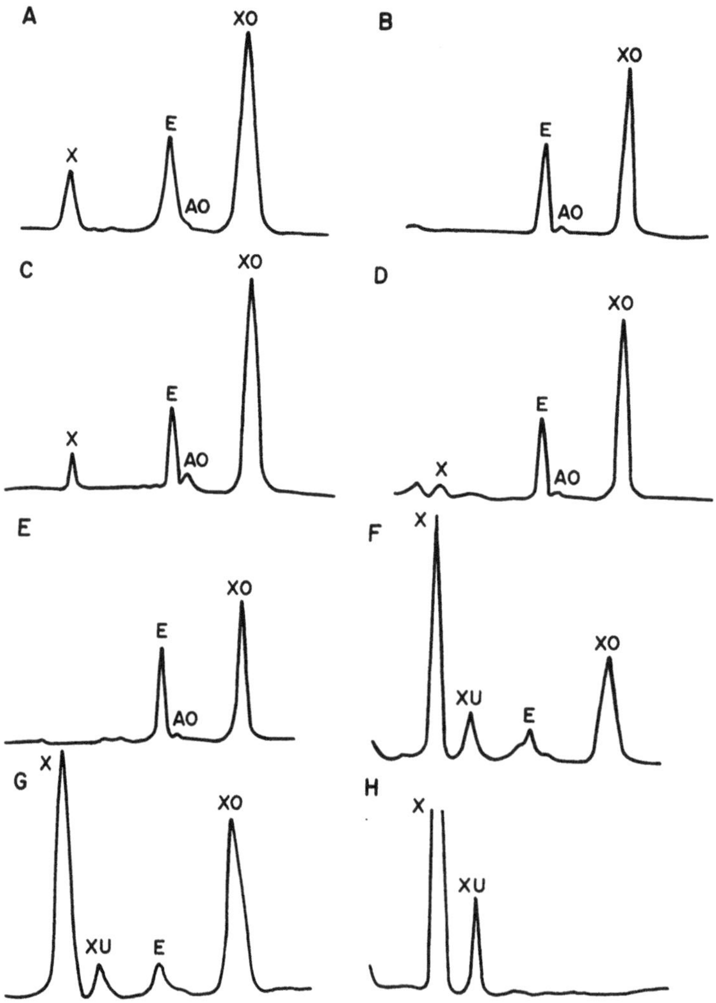

Fig. 4. Liquid chromatograms of D-xylose conversion by *Candida* sp. C2 and mutants in the presence of D-xylose isomerase. A, C2; B, C24; C, C210; D, C28; E, XF217; F, C232; G, C282; H, C26. X, D-xylose, XU, D-xylulose; E, ethanol; AO, D-arabitol; XO, xylitol. Incubation was carried out at 30 C for 36 h. D-xylose isomerase concentration was 30 g l^{-1} and the initial D-xylose concentration was 10% (w/v)

When xylose isomerase was present, slightly different metabolic product profiles were observed. In most of the yeast strains, D-xylose and D-xylulose were utilized by yeast simultaneously to produce mainly ethanol and xylitol with small amounts of D-arabitol also being produced (Fig. 4A-E). Mutant strains, C232 and C282, utilized D-xylulose only sparingly, again xylitol and ethanol were the products (Figs. 4F and G). Neither D-xylulose nor D-xylose was utilized by C26 under the incubation conditions (Fig. 4H).

3.3 Time-Course of D-Xylose Conversion by Selected Mutants

Figure 5 compares the time-course of D-xylose utilization and product formation by the wild-type strain and five mutants. The wild-type strain, C2, produced predominantly xylitol from D-xylose with small amounts of glycerol and ethanol also being produced (Fig. 5A) A slight increase in glycerol and ethanol production was observed in a mutant, C210 (Fig. 5B). Increased ethanol production at the expense of xylitol was observed in mutant C28 (Fig. 5C). A large increase in ethanol production at the expense of xylitol was observed in mutant, XF217 (Fig. 5D). There is an increase in D-arabitol production over that of the wild-type strain at the expense of xylitol in mutant C26 (Fig. 5E). And finally, a respiratory deficient mutant, 217-Pl, that exhibited a reduced D-xylose utilization rate was obtained from mutant strain XF217 (Fig. 5F).

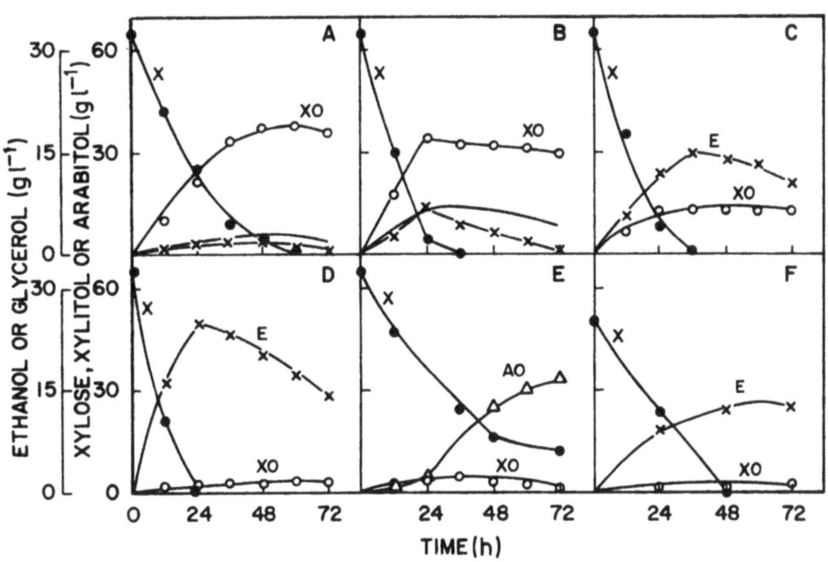

Fig. 5. Aerobic conversion of D-xylose by *Candida* sp. C2 and mutants. A, wild-type, C2; B, C210; C, C28; D, XF217; E, C26; F, 217-Pl. Incubation was carried out at 30 C with an initial pH of 5.6. The initial cell density was 1×10^8 cells per ml, and the intial D-xylose concentration was 65 g l^{-1}

4 Physiological Characterization of *Candida* sp. C2 and XF217

4.1 Carbohydrate Utilization

A comparison of carbohydrate utilization and product formation by wild-type, *Candida* sp. C2, and mutant XF217 on five different substrates was investigated by incubating yeasts under both aerobic and oxygen-limited fermentative conditions (Table 6). Both yeast strains utilized D-glucose readily under both incubation conditions to produce ethanol as the major product. When D-xylose was the substrate, strain C2 produced xylitol as the major product along with small amounts of ethanol, and D-arabitol. Strain XF217 produced mainly ethanol along with greatly reduced amounts of xylitol in comparison to strain C2. The rate of D-xylose utilization under fermentative conditions was half of that attained under aerobic conditions for both yeast strains.

D-Xylulose was a better substrate for both yeast strains than D-xylose. More ethanol was produced by C2 from D-xylulose than D-xylose although xylitol remained the major product. A high concentration of ethanol was produced from D-xylulose by XF217 and some xylitol was also produced. Smaller amounts of ethanol are

Table 6. Utilization and conversion of carbohydrates by *Candida* sp. C2 and XF217

Organism	Carbohydrates[a]	Fermentation[b] conditions	% Substrate consumed	Products %		
				ETOH	Arabitol	Xylitol
Candida sp. C2						
	D-Glucose	A	5.0	2.2	0	0
		N	5.0	2.2	0	0
	D-Xylose	A	3.05	0.3	0.1	2.28
		N	1.54	0.2	0.1	1.02
	D-Xylulose	A	5.0	0.78	0	2.48
		N	5.0	1.1	0	2.36
	Xylitol	A	0.4	0	0	—
		N	0.37	0	0	—
	L-Arabinose	A	1.42	0	0.94	0
		N	0.62	0	0.28	0
Candida XF217						
	D-Glucose	A	5	2.1	0	0
		N	5	2.2	0	0
	D-Xylose	A	4.27	1.41	0	0.64
		N	2.51	0.78	0	0.24
	D-Xylulose	A	5.0	1.57	0	0.16
		N	5.0	2.1	0	0.42
	Xylitol	A	0.21	0	0	—
		N	0.05	0	0	—
	L-Arabinose	A	2.36	0.1	1.27	0
		N	1.83	0.2	0.84	0

[a] Initial substrate concentration was 5% (w/v) and incubation time was 48 h
[b] A, aerobic; N, fermentative

produced under aerobic rather than under fermentative conditions due to the utilization of ethanol as a carbon and energy source by yeasts.

Xylitol is not a good substrate even though both yeasts produced xylitol readily from both D-xylose and D-xylulose. The possible lack of transport mechanism could account for this. L-Arabitol is the only product from L-arabinose by strain C2. Small amounts of ethanol were produced by XF217 in addition to the production of L-arabitol.

The major difference in strain C2 and XF217 is the ability of XF217 to produce higher amounts of ethanol from both D-xylose and D-xylulose. It is also interesting to note that small quantities of ethanol were produced from L-arabinose by XF217.

4.2 D-Xylose Reductase

D-Xylose reductase activity was measured spectrophotometrically by following the decrease in optical density at 340 nm due to the oxidation of NADPH when D-xylose was used as the substrate [23].

Both strains, C2 and XF217, exhibited good D-xylose reductase activity when D-xylose was the growth substrate. Only small enzyme activity was detected when D-glucose was the growth substrate (Fig. 6).

The substrate specificity of NADPH-linked polyol dehydrogenase activity was measured using either D-xylose, L-xylose, D-arabinose, L-arabinose, xylitol or D-glucose as the substrate. Results in Fig. 7 show that both D-xylose and L-arabinose are the substrates for this enzyme. No activity was observed with either L-xylose, D-arabinose or xylitol. Small amounts of activity were observed when D-glucose

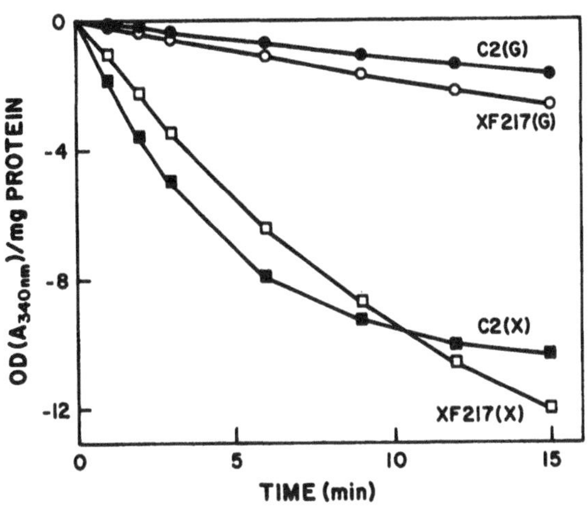

Fig. 6. Cell-free D-xylose reductase activities of *Candida* sp. C2 and XF217. G, D-glucose grown cells; X, D-xylose grown cells. The reaction mixture consisted of the following: 50 µmol of tris-hydrochloride (pH 7.5), 10 µmol of dithiothreitol, 0.1 µmol of NADPH and 0.1 mmol of D-xylose, adjusted to a volume of 1.3 ml with distilled water. The reactions were carried out in 1 cm light path curvetts and started by the addition of D-xylose. The rates of reduction of D-xylose were measured by following the change in absorbance at 340 nm using a Gilford Spectrophotometer

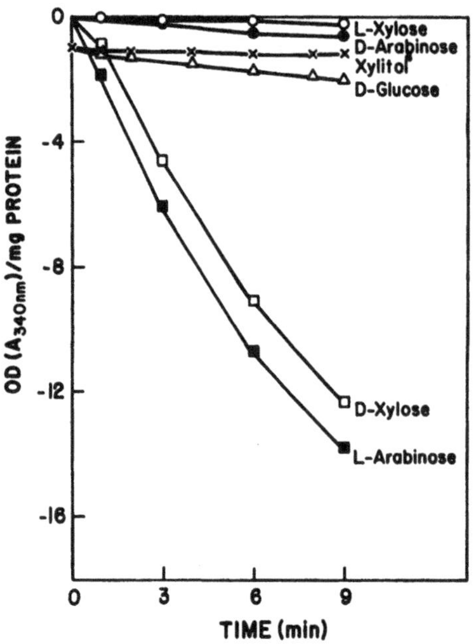

Fig. 7. Substrate specificity of D-xylose reductase from *Candida* sp. C2

was the substrate. Thus, the substrate specificity of this enzyme is similar to that demonstrated with enzymes isolated from other yeasts (see Table 2). The reverse reaction using NADP as co-factor was also examined using xylitol, D-arabitol, L-arabitol or sorbitol as the substrate. The results showed that xylitol is the substrate when NADP is the co-factor but not when NAD is used.

Co-factor specificity of D-xylose reductase was examined using D-xylose as the substrate. The results in Table 7 show that this enzyme is NADPH dependent and NADH cannot replace NADPH. Similar results were obtained when L-arabinose was used as a substrate (Fig. 8). Thus, both yeast strains exhibited good D-xylose reductase activities when D-xylose was the carbon and energy source for cell growth and this enzyme is similar to those reported in other yeasts.

Table 7. Co-Factor specificity for xylose reductase;
Substrate: D-Xylose; Reaction Time: 6 min

Co-factor	Relative activity (%)
Control (no co-factor)	0
$NADPH + H^+$	100
$NADH + H^+$	3
$NADP^+$	0
NAD^+	0

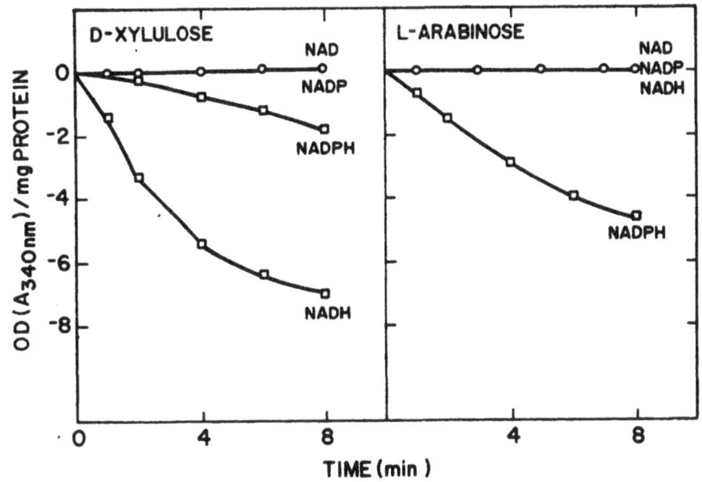

Fig. 8. Effect of co-factors on reactivity of D-xylulose and L-arabinose

4.3 Xylitol Dehydrogenase

Xylitol dehydrogenase activity was measured spectrophotometrically by following the increase in optical density at 340 nm due to the reduction of NAD when xylitol was used as the substrate [45].

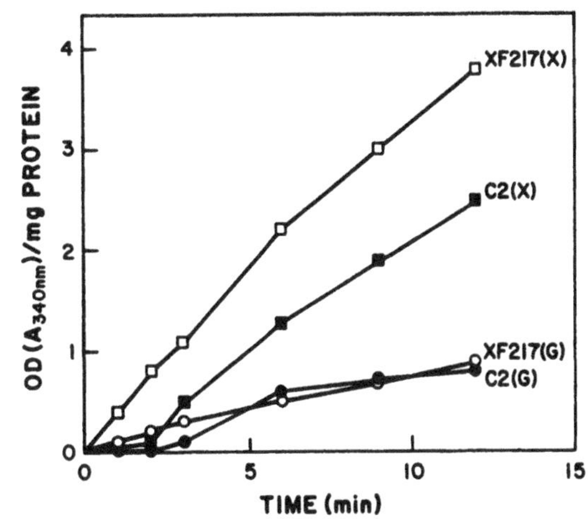

Fig. 9. Cell-free xylitol dehydrogenase activities of *Candida* sp. C2 and XF217. X, D-xylose grown cells; G, D-glucose grown cells. The reaction mixture consisted of the following: 50 μmol of tris-hydrochloride (pH 7.5), 10 μmol of $MgCl_2$, 5 μmol of dithiothreitol, 0.1 μmol of NAD^+ and 30 μmol of xylitol, adjusted to a volume of 1.3 ml with distilled water. The reactions were carried out in 1 cm light path curvettes and started by the addition of xylitol. The rates of oxidation of xylitol to D-xylulose were measured by following the change in absorbance at 340 nm using a Gilford Spectrophotometer

Cell-free extracts from both yeast strains exhibited good xylitol dehydrogenase activity when D-xylose was the growth substrate (Fig. 9). Lower enzyme activity was detected when D-glucose was the growth substrate. In XF217, the specific activity of this enzyme is 30% higher than that obtained from wild-type strain when D-xylose was the growth substrate.

Substrate specificity of NAD-linked polyol dehydrogenase activity was measured using either xylitol, D-arabitol or L-arabitol as the substrate. Results in Fig. 10 indicate that xylitol is the preferred substrate; small amounts of activity were observed with D-arabitol as a substrate but not with L-arabitol.

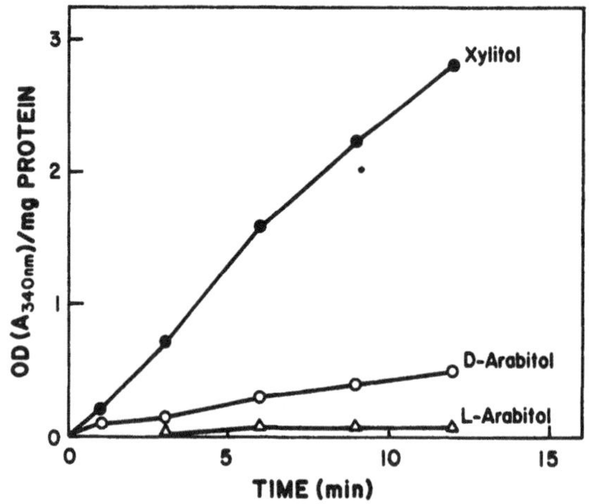

Fig. 10. Substrate specificity of xylitol dehydrogenase from *Candida* sp. XF217

Xylitol dehydrogenase activity required NAD as a co-factor. NADP could partially replace NAD as the co-factor when xylitol was used as the substrate (Table 8). When D-xylulose was used as the substrate, NADH was required and NADPH could partially replace it. This indicates that xylitol dehydrogenase can carry out the reduction of D-xylulose (Fig. 8).

Table 8. Co-factor specificity for xylitol dehydrogenase;
Substrate: Xylitol; Reaction time: 6 min

Co-factor	Relative activity (%)
Control (no co-factor)	0
NAD^+	100
$NADP^+$	12
$NAD^+ + H^+$	8
$NADP^+ + H^+$	0

4.4 D-Xylulokinase

D-Xylulokinase activity was measured spectrophotometrically by following the decrease in optical density at 340 nm due to the oxidation of NADH when D-xylulose was the substrate. An ATP regeneration system was also included [48]. The overall reaction to measure D-xylulokinase activity is as follows:

1) D-Xylulose + ATP → D-Xylulose-5-Phosphate + ADP
2) Phosphoenol pyruvate + ADP → Pyruvate + ATP
3) Pyruvate + NADH → L-Lactate + NAD

Cell-free extracts from XF217 exhibited high D-xylulokinase activities especially those from yeast grown in D-xylose. D-Xylulokinase activity in either D-glucose or D-xylose grown C2 cells is only 44% of the D-glucose grown XF217 cells and 17% of the D-xylose grown XF217 cells (Fig. 11).

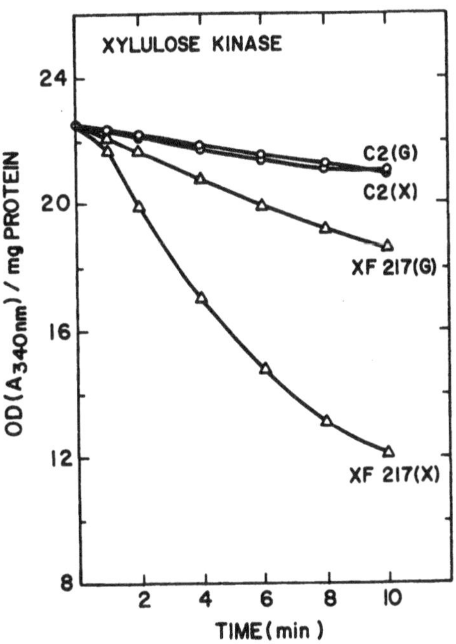

Fig. 11. Cell-free D-xylulokinase activities of *Candida* sp. C2 and XF217. G, D-glucose grown cells; X, D-xylose grown cells. The reaction mixture consisted of the following: 50 μmol of ATP, 5 μmol of $MgCl_2$, 50 μmol of KCl, 1 μmol of dithiothreitol, 1 μmol of EDTA, 0.1 μmol of NADH, 1 μmol of phosphoenolpyruvate, 0.3 μg per ml of rabbit muscle enzyme, and 1 μmol of D-xylulose, adjusted to a volume of 1.3 ml with distilled water. The reactions were carried out in 1 cm light path curvettes and started by the addition of D-xylulose reaction rates were measured by following the change in absorbance at 340 nm using a Gilford Spectrophotometer

4.5 D-Xylose Isomerase

D-Xylose isomerase catalyzes the interconversion of D-xylose to D-xylulose, D-glucose to D-fructose and other aldoses to ketoses [51,52]. This enzyme is found in many prokaryotes that utilize D-xylose as a carbon and energy source. It is believed that

D-xylose isomerase carries out the initial step of D-xylose metabolism in prokaryotes [53-55]. No conclusive evidence has demonstrated the existence of this enzyme in either yeasts or mycelial fungi. To measure the presence of D-xylose isomerase in yeasts, two types of reactions were carried out. One measures D-xylulose formation from D-xylose by liquid chromatography and the other measures D-glucose formation from D-fructose. As shown in Table 9, no D-xylose isomerase can be detected in D-xylose grown cells of either strains C2 or XF217. This enzyme activity has not been detected in cell-free extracts of many D-xylose grown yeasts or mycelial fungi.

Table 9. Xylose isomerase (Glucose isomerase) activity[a]

Organism	Xylose → Xylulose ($mg\ ml^{-1}$)			Fructose → Glucose ($mg\ ml^{-1}$)		
	35 C	50 C	70 C	35 C	50 C	70 C
Bacterium	1	7	16	0.8	6.2	30.4
Candida sp. C2	0	0	0	0	0	0
XF217	0	0	0	0	0	0

[a] The reaction mixture consisted of 50 μmol of α-D-glycerol phosphate at pH 7.0, 8.3 μmol or $MgCl_2$, 0.83 μmol of $CoCl_2$ and 2 mmole of D-fructose or 0.67 mmol of D-xylose, adjusted to a 2.5 ml volume with distilled water. The reaction was started by the addition of cell-free extract or whole-cell perparation, and stopped by the addition of trichloroacetic acid to a final concentration of 5%. After 1 h of incubation, the suspension was centrifuged at 8000 x g for 10 min, and the supernatant was measured for D-glucose formed from D-fructose and D-xylulose formed from D-xylose

5 Discussion

The results of the comparative study of enzymes involved in the early steps of D-xylose metabolism in *Candida* sp. and XF217 can be summarized as follows (Table 10):

1) No D-xylose isomerase activity can be detected in either yeast strain.
2) The presence of D-xylose is required for D-xylose reductase activity in both strains. The specific activity of this enzyme is similar in both strains. D-Xylose and L-arabinose are the substrates for this enzyme and the co-factor is NADPH.
3) Both yeast strains exhibit xylitol dehydrogenase activity especially when D-xylose is used as the growth substrate. The enzyme activity in XF217 is higher than that of the wild-type yeast. This enzyme is specific for xylitol with NAD as the co-factor. The reverse reaction was observed when D-xylulose was used as the substrate and NADH was the co-factor.
4) D-Xylulose kinase activity in XF217 was shown to be 3 times higher than in the wild-type yeasts. The presence of D-xylose is not essential for the appearance of this enzyme.

Table 10. D-xylose metabolic enzyme activities in *Candida* sp. C2 and XF217. Specific activity: 0.1 unit optical density ($A_{340\,nm}$) changed per minute reaction per mg protein at 30 C

Enzyme	Specific activity	
	C2	XF217
Xylose isomerase	34	0
Xylose reductase		24
Xylitol dehydrogenase	9	17
Xylulose kinase	4	15

The initial step of D-xylose metabolism in yeast for conversion of D-xylose to xylitol requires NADPH. This NADPH can be regenerated and supplied through the oxidation of D-glucose-6-phosphate to D-ribulose-5-phosphate (see Fig. 1). Many yeasts, especially *Candida* yeasts, are high in glucose-6-phosphate dehydrogenase and 6-phosphogluconic dehydrogenase [27,56,57]. They also show high transaldolase and transketolase activities that allow them to convert pentosephosphate to fructose phosphate and triosephosphate [58,59]. It is known that the oxidation of NADPH causes a stimulation of the pentose phosphate pathway in yeasts [56]. Thus, it is likely that the rate of D-xylose reduction and the activity of enzymes utilized within the pentose phosphate pathway are controlled by the availability of NADP and NADPH. The generation of NADP by reduction of D-xylose to xylitol may stimulate the activity of glucose-6-phosphate dehydrogenase. This, in turn, could stimulate the activation of the pentose phosphate pathway. In some yeast strains the oxidation of glucose-6-phosphate to provide a continous supply of NADPH and high activity of enzymes within the pentose phosphate pathway results in the continuous reduction of D-xylose to xylitol.

The increase in xylitol dehydrogenase and D-xylulokinase activities in ethanol-producing yeast mutants such as XF217 would shift the reaction toward D-xylulose formation instead of xylitol, thus, results in the further metabolism of D-xylulose-5-phosphate and ultimately yields ethanol.

The enhanced ability of the yeast mutant, C26, to produce D-arabitol from D-xylose is interesting in that D-ribulose-5-phosphate is the precusor of D-arabitol [60,61] and D-ribulose-5-phosphate can be derived from the oxidation of glucose-6-phosphate or D-xylulose-5-phosphate (see Fig. 1). The ability of yeast mutant, C26, to produce significant quantities of D-arabitol from D-xylose indicates that this yeast is high in phosphoketopentoepimerase, D-ribulose-5-phosphatase or arabitol dehydrogenase. The inability of strain C26 to utilize D-xylulose (Fig. 4H) is intriguing since D-xylulose, in general, is a better substrate than D-xylose for yeasts. Many yeast strains that cannot utilize D-xylose utilize D-xylulose readily [8,14,62]. Most strains examined utilized D-xylulose at a faster rate than D-xylose. A possible explanation for the unusual behavior of C26 in response to D-xylulose metabolism may be the lack of an uptake carrier. The response of C26 to D-xylulose indicates that yeasts possess different carriers for D-xylose and D-xylulose transport.

6 Conclusion

D-Xylose and L-arabinose are good substrates for yeasts and xylitol and L-arabitol are the common metabolic products. The two-step conversion of aldopentoses to ketopentoses in yeasts through an oxidation-reduction reaction is the major difference between eukaryotes and prokaryotes. Prokaryotes convert aldopentoses to ketopentoses through direct isomerization. Yeast mutants that exhibit modified product patterns can be obtained through mutation-selection. This provides a basis for selecting yeast strains that can produce ethanol from pentoses. The study of pentose metabolism in yeasts can provide valuable information that can lead to a better understanding of metabolic regulation. It can also provide the basis for improvement of yeast strains through genetic engineering to obtain better strains for industrial purposes.

7 Acknowledgment

We thank Mr. Allen Anderson and Ms. Cecilie Hunter for carrying out the analyses. Thanks are also due to Ms. Tanya Claypool and Ms. Ginny Garrison for assistance. We express our appreciation to Savannah Foods and Industries, and U.S. Sugar Corporation for sponsoring this research effort.

8 References

1. Aspinall, G. O.: Adv. Carbohydr. Chem. *14*, 429 (1959)
2. Siegel, S. M.: In: Int. Series of Monographs on Pure And Applied Biology. (Wareing, P. F., Galston, A. W. eds.), Vol. 2, The Plant Cell Wall
3. Chang, M. M., Chou, T. Y. C., Tsao, G. T.: Adv. Biochem. Engng. *20*, 15 (1981)
4. Schmidt, O., Walter, K.: Eur. J. Appl. Microbiol. *5*, 69 (1978)
5. Dekker, R. F. H., Richards, G. N.: Adv. Carbohydr. Chem. Biochem. *32*, 377 (1976)
6. Barnett, J. A.: ibid. *32*, 125 (1976)
7. Onishi, H., Suzuki, T.: Agr. Biol. Chem. *30*, 1139 (1966)
8. Gong, C. S., Claypool, T. A., McCracken, L. D., Maun, C. M., Ueng, P. P., Tsao, G. T.: Biotech. Bioeng. *25*, 85 (1983)
9. Barnett, J. A.: In: The Fungi. (Ainsworth, G. B., Sussman, A. S. eds.), Vol. III, p. 557, New York: Acad. Press 1968
10. Onishi, H.: Adv. Food Res. *12*, 53 (1963)
11. Lewis, D. H., Smith, D. C.: New Phytol. *66*, 143 (1967)
12. Lewis, D. H., Smith, D. C.: ibid. *66*, 185 (1967)
13. Brown, A. D.: Bacteriol. Rev. *40*, 803 (1976)
14. Gong, C. S., Chen, L. F., Flickinger, M. C., Chiang, L. F., Tsao, G. T.: Appl. Environ. Microbiol. *41*, 430 (1981)
15. Jeffries, T. W.: Biotechnol. Letters *3*, 213 (1981)
16. Schneider, H., Wang, P. Y., Chan, Y. K., Maleszka, R.: ibid. *3*, 89 (1981)
17. Slininger, P. J., Bothast, R. J., Van Cauwenberge, J. E., Kurtzman, C. P.: Biotech. Bioeng. *24*, 371 (1982)
18. Gong, C. S., McCracken, L. D., Tsao, G. T.: Biotechnol Lett. *3*, 245 (1981)
19. Gong, C. S., Chen, L. F., Flickinger, M. C., Tsao, G. T.: Adv. Biochem. Engng. *20*, 93 (1981)
20. Gong, C. S.: Annu. Rep. Ferm. Vol. 6 (in press)
21. Chiang, C., Sih, C. J., Knight, S. G.: Biochim. Biophys. Acta *29*, 664 (1958)

22. Chiang, C., Knight, S. G.: ibid. *35*, 454 (1959)
23. Chiang, C., Knight, S. G.: Biochem. Biophys. Res. Commu. *3*, 554 (1960)
24. Chiang, C., Knight, S. G.: Biochim. Biophys. Acta *46*, 271 (1961)
25. Chiang, C., Knight, S. G.: Nature (London) *188*, 79 (1960)
26. Veiga, L. A., Bacila, M., Horecker, B. L.: Biochem. Biophys. Res. Commu. *2*, 440 (1960)
27. Chakravorty, M., Veiga, L. A., Bacila, M., Horecker, B. L.: J. Biol. Chem. *237*, 1014 (1962)
28. Moret, V., Sperti, S.: Arch. Biochem. Biophys. *98*, 124 (1962)
29. Suzuki, T., Onishi, H.: Appl. Microbiol. *25*, 850 (1973)
30. Tomoyeda, M., Horitsu, H.: Agr. Biol. Chem. *28*, 139 (1964)
31. Höfer, M., Botz, A., Kotyk, A.: Biochim. Biophys. Acta. *252*, 1 (1971)
32. Veiga, L. A.: J. Gen. Appl. Microbiol. *14*, 65 (1968)
33. Horitzu, H., Tomoeda, M., Kumagai, K.: Agr. Biol. Chem. *32*, 514 (1968)
34. Birken, S., Pisano, M. A.: J. Bact. *125*, 225 (1976)
35. Clancy, F. G., Coffey, M. D.: J. Gen. Microbiol. *120*, 85 (1980)
36. Bertrand, G.: Compt. Rend. Acad. Sci. *126*, 762 (1898). In: Agr. Biol. Chem. *30*, 962 (1966)
37. Blakley, R. L.: Biochem. J. *49*, 257 (1951)
38. McCorkindale, J., Edson, N. L.: ibid. *57*, 518 (1954)
39. Hollman, S.: In: Int. Symp. on Metabolism Physiology, and Clinical Use of Pentoses and Pentitols. (Horecker, B. L., Lory, K., Takayi, Y. eds.), p. 50, Berlin: Springer 1967
40. Mortlock, R. P.: Catabolism of Unnatural Carbohydrates by Microorganisms. In: Advances in Microbial Physiology. (Rose, A. H., Tempest, D. W. eds.), Vol. 13, p. 1, New York: Academic Press 1976
41. Arcus, A. C., Edson, N. L.: Biochem. J. *64*, 385 (1956)
42. Washüttl, J., Riederer, P., Bancher, E.: J. Food Sci. *38*, 1262 (1973)
43. Mäkinen, K. K., Söderling, E. J.: ibid. *45*, 367 (1980)
44. Horitsu, H., Tomoeda, M.: Agr. Biol. Chem. *30*, 962 (1966)
45. Veiga, L. A.: J. Gen. Appl. Microbiol. *14*, 79 (1968)
46. Sugai, J. K., Veiga, L. A.: An. Acad. Brasil, Cienc. *53*, 183 (1981)
47. Stumpf, P. K., Horecker, B. L.: J. Biol. Chem. *218*, 753 (1956)
48. Simpson, F. J.: In: Methods in Enzymology (ed. Wood, W. A.) *9*, 454 (1966)
49. Ladisch, M. R., Tsao, G. T.: J. Chromatogr. *166*, 35 (1978)
50. Ladisch, M. R., Anderson, A. W., Tsao, G. T.: J. Liq. Chromatogr. *2*, 745 (1979)
51. Yamanaka, K.: Biochim. Biophys. Acta *151*, 670 (1968)
52. Sanchez, S., Smiley, K. L.: Appl. Microbiol. *29*, 745 (1975)
53. Mortlock, R. P., Wood, W. A.: J. Bacteriol. *88*, 838 (1964)
54. Wood, W. A.: Annu. Rev. Biochem. *35*, 521 (1966)
55. Shamanna, D. K., Sanderson, K. E.: J. Bacteriol. *139*, 64 (1979)
56. Osmond, C. B., ApRees, T.: Biochim. Biophys. Acta *184*, 35 (1969)
57. Lobo, Z., Maitra, P. K.: Mol. Gen. Genet. *185*, 367 (1982)
58. Höfer, M. K., Brand, K., Decker, K., Becker, J. U.: Biochem. J. *123*, 855 (1971)
59. Sols, A., Gancedo, C., DelaFuente, G.: In: The Yeasts. (Rose, A. H., Harrison, J. S. eds.), Vol. 2, p. 271, New York: Academic Press 1971
60. Spencer, J. F. T., Neish, A. C., Blackwood, A. C., Sallans, H. R.: Can. J. Biochem. Physiol. *34*, 495 (1900)
61. Ingram, J. M., Wood, W. A.: J. Bacteriol. *39*, 1186 (1965)
62. Wang, P. Y., Schneider, H.: Can. J. Microbiol. *26*, 1165 (1980)

Ethanol Production from D-Xylose and Several Other Carbohydrates by *Pachysolen tannophilus* and Other Yeasts

Henry Schneider, R. Maleszka, L. Neirinck, I. A. Veliky, P. Y. Wang and Y. K. Chan
Molecular Genetics Group, Division of Biological Sciences, National Research Council of Canada*, Ottawa, Canada K1A OR6

Yeasts had been considered as unable to produce ethanol from pentoses. Ethanol from such sugars is of interest because of the enhancement which would thereby follow in the economics of the bioconversion of lignocellulosics. In one approach to obtain yeasts which convert D-xylose to ethanol screening was carried out. Several were identified and some of the experimental parameters favoring the conversion were investigated. With one of the better converters, *Pachysolen tannophilus*, ethanol could be obtained from four of the five major phytomass sugars; D-glucose, D-mannose, D-galactose and D-xylose. Using a mutant of *P. tannophilus* selected for more rapid growth on D-galactose than the wild type, yields from mixtures containing these sugars could be high, in the 83–90% range. Yeasts were also shown to be able to produce ethanol from glycerol and from mixtures of D-cellobiose plus D-xylose, materials whose conversion to ethanol is of potential economic importance. In another approach to obtain suitable yeasts, a gene encoding for D-xylose isomerase in *Escherichia coli* was cloned. Appropriate introduction of this gene into a suitable yeast host is expected to allow the transformant to convert D-xylose into ethanol.

1 Introduction

Obtaining efficient conversion of five carbon sugars to ethanol has been a longstanding problem in the bioconversion of lignocellulosics. Although many yeasts efficiently convert six carbon sugars to ethanol, and many can also utilize pentoses oxidatively, they have generally been considered unable to produce ethanol from pentoses [1,2].

*NRCC Publication No. 20638

There are a number of approaches to obtain yeasts which could efficiently produce ethanol from pentoses as well as from hexoses. One involves a search for naturally occurring organisms with the desired property. Another centers on the use of genetic engineering methods to introduce into an appropriate yeast host, genes that would permit the efficient production of ethanol from D-xylose. The present paper reviews recent progress in our laboratory using these two approaches.

2 Naturally Occurring Yeasts that Convert D-Xylose into Ethanol

2.1 *Pachysolen tannophilus*

Recently, the yeast *P. tannophilus* was identified as being able to convert D-xylose into ethanol [3,4]. Species of *Candida* capable of carrying out the conversion were

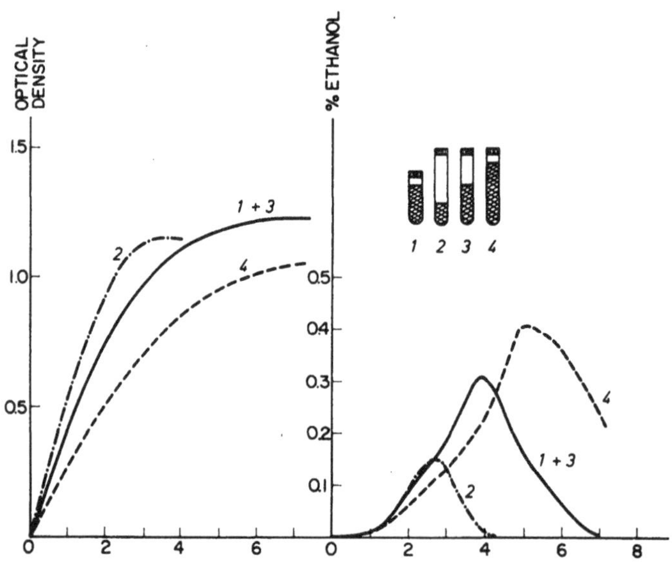

Fig. 1. Effect of extent of aeration on growth and ethanol production by *Pachysolen tannophilus* NRRLY 2460 on D-xylose.
Growth (left hand side) and ethanol production behavior (right hand side) varied with the extent of aeration. As aeration decreased, growth rate decreased. Concomitantly, the maximum concentration of medium ethanol and the time to reach this maximum increased. The variation in growth rate with aeration is the result of differences in the extent of oxygen limitation. The medium was 2% D-xylose in 0.67% yeast nitrogen base.
Aeration was varied by using different volumes of medium in screw-capped test tubes, 16 mm in diameter and either 125 or 75 mm in length. They were loosely capped in an identical manner and rotated at 60 rpm about their long axis, kept at 60° from the vertical, at 30 °C [3]. The volumes of medium employed were 10 ml for tubes 1 and 3, 6 ml for tube 2, and 12 ml for tube 4.
Tubes 1 and 3 differed only in the volume of the air space they contained. Since they yielded similar results, the amount of air available to the cultures depends on its diffusion into the tube at the cap-tube junction

identified recently as well [5,6]. *P. tannophilus* requires oxygen for growth, while alcohol production can occur either aerobically or anaerobically [3,4,7]. It is about one-third the size of *Saccharomyces cerevisiae* and is homothallic.

Ethanol production by *P. tannophilus* growing aerobically is characterized by dependance on the extent of aeration [3]. Some of the phenomena observed are depicted in Fig. 1. As aeration decreases growth rate decreases. Concomitantly, the peak value of ethanol in the medium increases, as well as the time to attain this maximum. With highly aerated cultures, it is possible to obtain only traces of ethanol, insert <0.1 %. An example is when 50 ml of 2 % D-xylose in 0.67 % yeast nitrogen base is inoculated to an optical density of 0.03, and then shaken at 150 rpm on a gyrotory shaker at 30 C in a 500 ml Erlenmeyer flask sealed with a porous plug.

Ethanol production by growing cells can be slow, is days being required for the peak value to appear (e.g. Fig. 1). The low production rate is attributed, in part, to the relatively long doubling time under the conditions used, estimated as 4 to 6 h. The appearance of appreciable amounts of ethanol in the medium also lags growth by about one day. The eventual decrease in ethanol level is due to its oxidation.

The use of cell recycling at high cell densities results in considerable improvement in the rate and yield of ethanol production [7]. With D-xylose as sole sugar source at the 2 % level, yields of 0.395 g ethanol per g D-xylose were obtained within 24 h at 37 C under aerobic conditions (Table 1). The yield at 30 C was somewhat lower, 0.360 g ethanol per g D-xylose, because the rate of production was smaller. Less ethanol was produced under anaerobic than under aerobic conditions at both 30 and 37 C. The yield of 0.395 g ethanol per g D-xylose corresponds to 78 % of theoretical

Table 1. Ethanol yield in 24 h by *Pachysolen tannophilus* NRRL Y 2460 recycled under different conditions*

Temperature (C)	Ethanol yield (g ethanol per g D-xylose)	
	Aerobically	Anaerobically
30	0.360	0.345
37	0.395	0.360

* The cells used had previously been recycled four times on 2 % D-xylose in 0.67 % yeast nitrogen base at 30 C under aerobic conditions. They were then recycled in fresh medium under aerobic and anaerobic conditions at both 30 and 37 C [7]

assuming that a) all of the D-xylose was consumed and b) that the D-xylose is converted to D-xylulose-5-phophate, then to glucose-6-phosphate, after which the normal glycolytic pathway is followed. The stoichiometry for this model is

$$3\ C_5H_{10}O_5 \rightarrow 5\ C_2H_5OH + 5\ CO_2$$

The improvement in the rate of ethanol production as the result of cell recycling is due, in part, to the high cell density, which provides a large number of cells capable of carrying out the conversion. The high cell density is also considered to play a role by keeping the level of dissolved oxygen suitably low. By choosing a suitable interval for recycling, it was also possible to minimize alcohol loss due to oxidation. Other factors which might play a role are the removal of inhibitory products, or the replenishment of trace nutrients, as the results of regular replacement of medium.

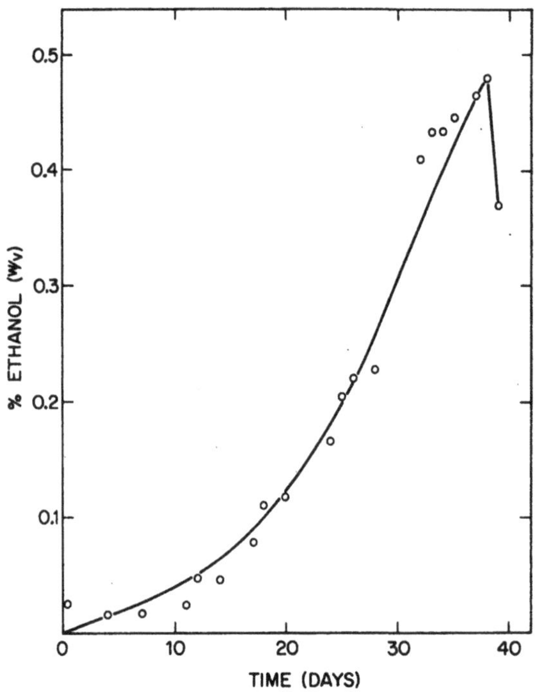

Fig. 2. Ethanol production by immobilized *Pachysolen tannophilus*. Strain NRRL Y2460 was immobilized in an alginate gel [8]. The bead diameter was 3 mm, the dimensions of the working volume of the column reactor were 34 mm in diameter and 210 mm in length, and the average flow rate was 79 ml per 24 h. The composition of the medium was, per 1, 10 g D-xylose, 1 g $(NH_4)_2SO_4$, 1 g $MgSO_4 \cdot 7H_2O$, 2.94 g $CaCl_2$, 1.0 g yeast nitrogen base, and trace elements [9]. The pH was adjusted to 5 with NaOH/HCl prior to autoclaving for 15 min. The temperature was 30 C

An enhancement in the rate of ethanol production was also obtained using immobilized cells [7]. In a static test system, where the beads were resuspended in fresh medium every 24 h, yields of 0.28 g ethanol per g D-xylose could be obtained from 2% D-xylose. A feature of the beads used was that productivity increased with each resuspension, attaining the value quoted above after nine days, at which time the experiment was terminated. Essentially similar results were obtained with flasks kept in an atmosphere of H_2 and CO_2 (Gas-Pak Jar) or in air. In subsequent, longer term experiments with a column of beads subjected to continuous flow of medium, a yield exceeding 90% of theoretical could eventually be attained (Fig. 2). This high yield is indicative of the potential of *P. tannophilus* in converting D-xylose to ethanol. The reason for the increase in productivity with time has still to be elucidated.

2.2 Growth of *P. tannophilus* under Anoxic Conditions

P. tannophilus does not grow anoxically on D-xylose [4]. Some species of *Saccharomyces* also do not grow anoxically on D-glucose. However, the latter will grow if the medium is supplemented with unsaturated lipids (oleic acid and ergosterol). The lack of anoxic growth of the *Saccharomyces* strains has been attributed to a requirement for oxygen in the synthesis of some unsaturated lipids. Such supplements were found to be ineffective with *P. tannophilus* on D-xylose [12]. Their ineffectiveness indicates that the need for oxygen for growth of *P. tannophilus* differs from that of the *Saccharomyces* species.

2.3 Ethanol from Sugar Mixtures Using *P. tannophilus*

D-xylose is only one of several sugars present in commercial substrates. Therefore, it is desirable to obtain organisms which can produce ethanol from as many of the component sugars as possible. Spent sulfite liquor was chosen as a model industrial feedstock in evaluating *P. tannophilus*. Two of the factors leading to this choice were a) the availability of the material in large amounts and (b) its appreciable content of D-xylose. An example of the composition of a softwood liquor (spruce) is 12% D-glucose, 51% D-mannose, 21% D-xylose, 12% D-galactose, 4% L-arabinose [10].

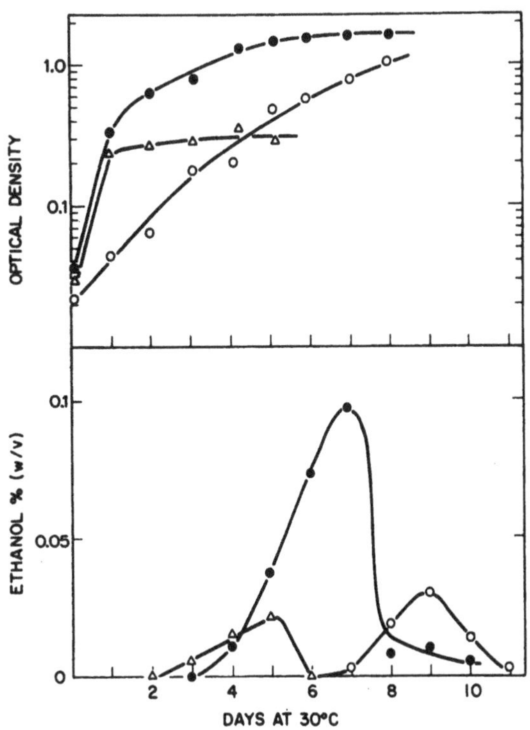

Fig. 3. Growth and Ethanol production behavior of *Pachysolen tannophilus* on D-galactose.
The upper portion depicts growth and the lower ethanol production. The media were 2% D-galactose (○), 0.1% D-glucose (△), and 2% D-galactose plus 0.1% D-glucose (●) in 0.67% yeast nitrogen base. Temperature was 30 C. The experiment was carried out aerobically in rotated tubes [3]

P. tannophilus converts D-glucose into ethanol [12]. It also readily produced ethanol from mannose [12], in agreement with the generalization that yeasts which produce ethanol from glucose also do so from mannose [2]. However, it is classified taxonomically as unable to convert D-galactose to ethanol, while being able to assimilate this hexose oxidatively [11]. Many other yeasts are similarly classified, but some can also convert D-galactose to ethanol. L-arabinose has been categorized as one of the sugars which yeasts, in general, are unable to convert to ethanol [2].

Prior to investigating sugar mixtures, the ability of *P. tannophilus* to produce ethanol from D-galactose was reinvestigated [12]. Initial experiments employed the aerobic batch conditions successful in getting ethanol from D-xylose [3]. Small amounts were produced from 2% D-galactose, 0.03% (Fig. 3, lower portion), a quantity which was much less than that from D-xylose, 0.53%. Ethanol became evident in the medium after a lag of seven days, a time when stationary phase was being approached (Fig. 3, upper portion). A higher concentration was formed, 0.097%, as the result of including 0.1% D-glucose in the medium. The increase came primarily from the D-galactose. The theoretical maximum from 0.1% D-glucose is 0.051% while the maximum found experimentally with 0.1% D-glucose alone was 0.02% (Fig. 3, lower portion). Although ethanol appeared earlier when glucose was present, after 3 instead of 7 days, it also appeared when stationary phase was being approached. Comparison of the growth curves on glucose and on D-glucose plus D-galactose suggests that in the mixture growth takes place first on glucose, because in both instances there is a burst of growth in the first 24 h.

Table 2. Ethanol from 2% D-galactose by recycled *Pachysolen tannophilus*

Cycle number	Optical density	Ethanol (% w/v) (24 h cycle)
1	0.07	0.02
2	0.14	0.19
3	0.50	0.33
4	0.85	0.40
5	1.00	0.40
6	1.10	0.55
7	1.20	0.60
8	1.40	0.66
9	1.60	0.67
10	2.00	0.70
11	>2.00	0.75
12	>2.00	0.76

The production of ethanol when stationary phase was being approached suggested that high cell density favored the conversion, hence that cell recycling would be beneficial. Such was found to be the case (Table 2) [12]. The ethanol concentration increased with cycle number, concomitantly with optical density. The highest concentration, 0.76% was obtained after the 12th cycle. During this cycle more than 95% of the D-galactose had been consumed. The alcohol concentration of 0.76%

corresponds to a production of 1.49 moles per mole of D-galactose consumed, and represents a yield of 75% of theoretical, assuming that the D-galactose is converted to D-glucose 6-phosphate, after which the normal glycolytic pathway is followed.

Although wild-type *P. tannophilus* could efficiently produce ethanol from D-galactose when present as sole sugar source, the situation was more complex in mixtures simulating spent sulfite liquor [13]. In cell recycle experiments appreciable amounts of D-galactose remained after 24 h, up to 20% of that originally present, while more than 95% of the D-glucose, D-mannose and D-xylose had been utilized. The problem was attributed to the relatively low rate of utilization of D-galactose, as judged by the doubling time of ~24 h on 2% D-galactose. Accordingly, a mutant was isolated which grew faster on D-galactose as sole sugar source. The doubling time of that chosen for detailed study was 6.5 h on 2% D-galactose. With this mutant, D-galactose and all of the other sugars except L-arabinose were utilized completely within 18 h. Ethanol yields were in the range of 83–90%, depending on the composition of the mixture, assuming that D-glucose, D-mannose, D-xylose and D-galactose were convertible to ethanol, but not L-arabinose.

A small, but significant loss in yield was caused by the formation of products other than ethanol; acetate, xylitol, and arabinitol [13]. The loss was estimated as 5–8.3% for the mutant, depending on the composition of the mixture employed.

The results with mixtures show that *P. tannophilus* has the potential for use with a wide variety of commercial substrates. The four sugars which could be converted to ethanol, D-glucose, D-mannose, D-xylose, D-galactose, comprise in general more than 90% of those in phytomass. While ethanol production from the fifth common component, L-arabinose, would be desirable, it is generally the least abundant.

2.4 Other Yeasts that Convert D-xylose into Ethanol

While *P. tannophilus* has the potential to carry out the desired conversion with phytomass sugars, several factors require attention in considering use of the wild-type for industrial purposes.

a) It grows slowly on D-xylose under the aerobic conditions which lead to alcohol production (generation time >4 h).

b) In the presence of air, when D-xylose is the sole sugar source, it stops growing at an ethanol concentration greater than 4–5% (Fig. 4).

c) When D-xylose is the carbon source in relatively high concentration, an appreciable amount of xylitol can be formed. With 4% D-xylose in a batch, aerobic culture, ~40% of the D-xylose present initially can be converted to xylitol (Fig. 5). Although some of the xylitol is subsequently utilized, ethanol is not produced thereby.

d) Other by-products can be formed, acetate and arabinitol [13], which result in a decrease in ethanol yield.

As part of an effort to find strains with improved properties, a preliminary survey was made of the distribution among yeasts of the ability to convert D-xylose to ethanol during aerobic growth. Fifteen yeasts from seven genera were tested [14]. The data indicate that the ability is probably common, since it was found in six of of the seven tested (Table 3). The data indicate also that the ability is strain specific. There were differences in ethanol productivity between species in a genus

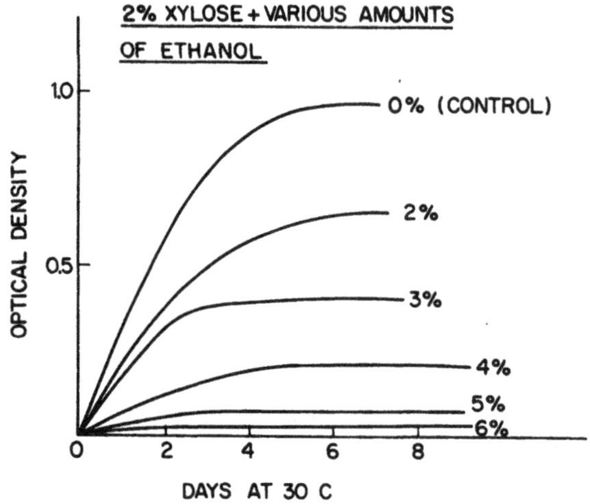

Fig. 4. The effect of ethanol concentration on the growth of *Pachysolen tannophilus* NRRL Y2460.
The medium contained 0.67% yeast nitrogen base plus 2% D-xylose and the concentrations of ethanol indicated. Growth was under aerobic conditions in rotated tubes (16 × 125 mm diameter, medium volume = 10 ml)

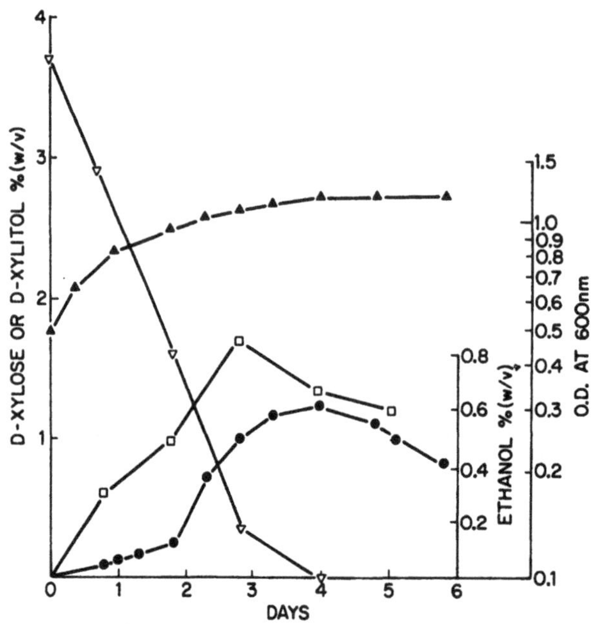

Fig. 5. The production of xylitol and ethanol from D-xylose by *Pachysolen tannophilus* NRRL Y2460.
As D-xylose concentration (∇) decreases and growth (▲) occurs, the concentration of xylitol (□) as well as of ethanol (●) increases, and then decreases. The eventual decrease in xylitol and ethanol is due to their utilization.
The volume of the bioreactor (Multigen, New Brunswick, New Brunswick, N.J.) was 1 L and contained 500 ml of medium consisting of 4% D-xylose in 0.67% yeast nitrogen base. The airflow rate was 50 ml min^{-1} and the temperature was 30 C

(e.g. *Hansenula anomala*, 0.03% from 4% D-xylose versus 0.23% from *Hansenula holstii*). There were also differences within species (e.g. *Candida tropicalis* ATCC 1369, 0.05% versus *C. tropicalis* ATCC 9968, 0.15%). In addition, there were medium effects. A rich medium enhanced ethanol production with some strains, but not others. For example, *P. tannophilus* was essentially unaffected by medium, while there was a manyfold change with *Metschnikowia pulcherrima*, from 0 to 0.23%.

Table 3. Ethanol production from D-xylose by various yeasts[a]

Yeast		Ethanol concentration % w/v		
		Anoxic[a]	Aerobic[a]	
		D-xylose	D-xylose	
		RM[b]	RM[b]	DM[c]
Candida guilliermondii	ATCC 22017	0.001	0.45	0.41
Candida terebra	ATCC 20022	0.035	0.30	0.25
Candida tropicalis	ATCC 9968	0.036	0.15	0.085
Candida tropicalis	ATCC 1369	0.046	0.050	0.005
Candida utilis	ATCC 9256	0.043	0.030	0
Debaryomyces polymorpha	NRRL Y-2022	0.019	0.20	0.025
Hansenula anomala	NRRL Y-30	0.001	0.030	0
Hansenula holstii	NRRL Y-2448	0.009	0.23	0.013
Hansenula saturnus	ATCC 9847	0.035	0.010	0
Kluyveromyces fragilis	ATCC 34439	0.055	0.050	0.020
Metschnikowia pulcherrima	NRRL Y-987	0.043	0.23	0
Pachysolen tannophilus	NRRL Y-2460	0.067	0.96	0.97
Pichia angophorae	NRRL Y-7118	0.046	0.26	0.18
Pichia etchellsii	NRRL Y-7121	0	0.24	0.011
Pichia guilliermondii	NRRL Y-2067	0	0.25	0.027

[a] Measured in rotated tubes [14];
[b] RM = rich medium; 1% yeast extract, 0.35% yeast nitrogen base, 0.1% casamino acids, 4% sugar;
[c] DM = defined medium; 0.67% yeast nitrogen base, 4% sugar. A value of zero denotes alcohol concentrations below 0.001%, the lowest value measureable under the conditions used

Because of the relatively small number of species surveyed, it is still unclear if taxonomy will be of great importance in finding strains of interest. Conversely, however, the strain specificity for the conversion of D-xylose to ethanol might be useful in taxonomic classification. The ability of some strains to produce ethanol from D-cellobiose and from glycerol (below) could also be useful in this regard.

Although the strains tested could produce some ethanol under the anoxic conditions employed, higher concentrations were obtained when air had access to the medium (Table 3). Also, none could produce ethanol from xylitol [14].

3 Other "Novel" Conversions of Carbohydrates to Ethanol by Yeasts

D-xylose is only one of several sugars which had been considered as nonconvertible to ethanol [2]. The success in getting the conversion with D-xylose through the use of aerobic conditions prompted determination if such conditions would be applicable with some of these other sugars. One investigated in this respect was D-galactose when it served as a carbon source for *P. tannophilus* (Sect. 2.3). Two others with which success has been obtained, and whose conversion to ethanol is of potential economic significance, are D-cellobiose and glycerol.

Alcohol production from D-cellobiose would be of particular interest if the organism involved could carry out the conversion with D-xylose as well. An organism with these properties could have advantages in the combined saccharification and production of ethanol from cellulosic solids. These would be

a) an increase in the rate of cellulose hydrolysis through obviation of the inhibition of cellulose activity by D-cellobiose accumulating in the culture and

b) the ability to process simultaneously the cellulose and hemicellulose fractions.

The required xylanases for hemicellulose hydrolysis can be produced along with cellulases by some cellulolytic fungi [15].

Very few yeasts which convert D-cellobiose to ethanol have been identified: 8 positive and 8 variable out of 439 species [11]. Although more than 100 assimilate the disaccharide, their inability to convert it to ethanol has not been established.

Table 4. Ethanol production by *Candida lusitaniae* ATCC 34449 on 2% D-xylose, 2% cellobiose and 2 + 2% mixtures of these sugars under various conditions

	Alcohol concentration (% w/v)		
Sugar	Aerobic growth [a]	Aerobic recycle [b]	Anoxic [c]
D-xylose	0.07	0.07	0.02
D-cellobiose	0.28	0.82	0.78
Mixture	0.35	1.20	0.80

[a] Peak value obtained using rotated tubes;
[b] At end of the fifth 24 h cycle;
[c] After 3 d, using the dense suspension of cells obtained by sealing the tubes used in the recycle experiment at the beginning of the ninth cycle

In addition, none of these organisms have been identified as being able to carry out the conversion with D-xylose.

In a search for organisms which produce ethanol from both sugars, striking results were obtained with *Candida lusitaniae* (Table 4)[16]. In mixtures of D-xylose and D-cellobiose, the amount of alcohol which it produced exceeded the sum of that when the sugars were tested individually. For example, in cell recycle experiments, the alcohol concentration after 24 h was 0.07% from 2% D-xylose, 0.82% from 2% D-cellobiose, but 1.2% from a 2% + 2% mixture of these sugars. Thus, one sugar stimulated the turnover of the other, or there was mutual stimulation. In addition the data were consistent with alcohol production occurring from both sugars, if the D-cellobiose catabolic pathway involves its conversion to D-glucose and subsequent processing via the normal glycolytic pathway.

Alcohol production from glycerol is of interest, because this polyol occurs as a by-product in the conversion of sugars to alcohol by *Saccharomyces* species[17]. The availability of a yeast that converts glycerol to ethanol could be used to improve ethanol yield, either by substitution for *Saccharomyces* species, or through concurrent or sequential processing. Glycerol has been considered as being used only oxidatively by yeasts[2], but transient ethanol formation by *C. utilis* was reported recently[18].

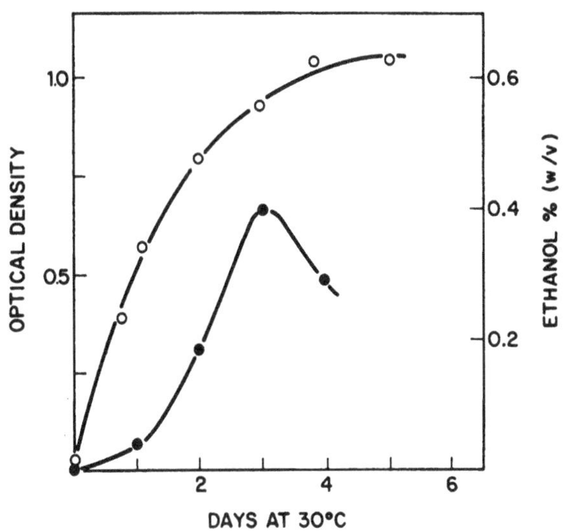

Fig. 6. Ethanol production from glycerol by *Pachysolen tannophilus* grown aerobically. The medium contained 2% glycerol in 0.67% yeast nitrogen base; ○, optical density; ●, ethanol concentration. Aerobic growth was carried out in rotated tubes at 30 C)[12]

P. tannophilus produced ethanol from glycerol when grown aerobically in batch culture (Fig. 6)[12]. The peak value from 2% glycerol corresponds to the production of 0.4 moles of ethanol per mole of glycerol consumed. Recycling decreased the amount of alcohol formed, in contrast to the behavior of D-xylose and D-galactose. After 24 and 48 h cycles, the values obtained after the ninth cycle were, respectively

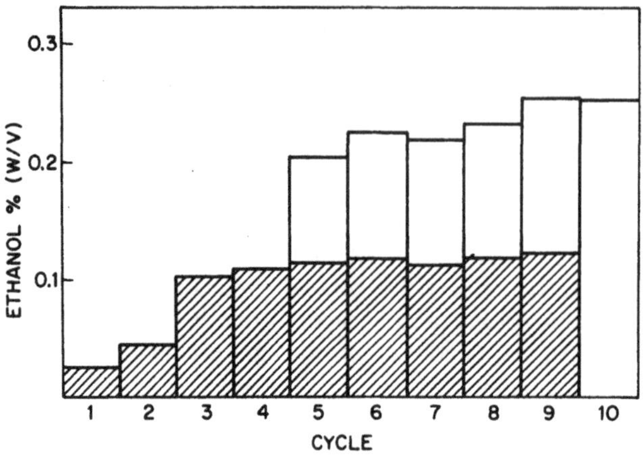

Figure 7. Ethanol production from gycerol by *Pachysolen tannophilus* during aerobic recycling: hatched portion, 24 h cycle; clear portions, 48 h cycle [12]

0.13 and 0.25%, (Fig. 7) in contrast to the peak value of 0.4% obtained in batch culture.

Trichosporon fermentans was also found to be able to convert glycerol to ethanol (Table 5). Thus, the ability to carry out the conversion is not uncommon among yeasts.

Table 5. Ethanol production by *Trichosporon fermentans* ATCC 10075 from various sugars under different culture conditions

Sugar[a]	Ethanol (% w/v)	
	Aerobic growth [b]	Aerobic recycle [c]
D-Glucose	0.17	0.71
D-Mannose	0.21	0.59
D-Galactose	0.12	0.31
D-Xylose	0.08	0.59
D-Cellobiose	0.02	0.27
Glycerol	0.14	0.06

[a] 2% in 0.67% yeast nitrogen base in all cases;
[b] Peak value in rotated tubes;
[c] At the end of the third 48 h cycle

An interesting facet of the physiology of *T. fermentans* is that in addition to being able to produce ethanol from glycerol, D-xylose, D-glucose, D-mannose and D-galactose, it can also degrade lignosulfonate [19].

4 Dependance of Ethanol Production on Experimental Conditions

Ethanol production by yeasts from the substrates investigated depends sensitively on experimental conditions. The extent of aeration is one of the more important of such conditions with *P. tannophilus* (Sect. 2.1) and with *Candida tropicalis* [5] on D-xylose. Aeration effects may also play a role in the enhancement of ethanol production by *P. tannophilus* on D-xylose when cell recyle methods are used, although other factors are likely to be involved as well (Sect. 2.1).

Some of the effects of experimental conditions found with the other substrates investigated are summarized in Table 5 using *T. fermentans* as an example. Recycling enhanced ethanol production from D-galactose and D-cellobiose, but decreased it from glycerol. With this organism there was also an effect of conditions on ethanol production from D-glucose and D-mannose, recycling yielding more than batch culturing.

Elucidation of the factors responsible for the sensitivity may be a necessary step in optimizing process conditions for converting D-xylose and other sugars to ethanol. One such factor has been indentified for *P. tannophilus* as the consumption of ethanol, at a rate depending on aeration level, while appreciable amounts of D-xylose or other sugars are still present in the medium [20]. A similar effect occurs with *C. tropicalis* ATCC 750.

The dependence of ethanol production from D-xylose on experimental conditions, coupled with differences in productivity between strains of the same species (Sect. 2.4), may have been a factor in the lack of earlier, general recognition that yeasts can convert D-xylose to ethanol, despite such claims [21,22].

5 Towards the Construction of Yeasts that Convert D-Xylose into Ethanol Using Genetic Engineering Methods

Yeasts of the genus *Saccharomyces* used industrially, e.g. brewers yeast, have commonly been considered as unable to produce ethanol from D-xylose. Providing them with this ability would be advantageous, because it should then be possible to employ the widespread technology which already exists for their use. Genetic engineering offers the possibility of constructing the desired type of yeast. A strategy we are using is based on the finding that several *Saccharomyces* species produce ethanol from D-xylulose [23-25]. Thus, these species would have the biochemical capability of producing ethanol from D-xylose, if they were transformed with a gene coding for D-xylose isomerase, which could also be suitably expressed in yeast. Genes for D-xylose isomerase exist in many procaryotes. The first step of the strategy has been completed with the isolation and characterization of an *Escherichia coli* plasmid bearing the *E. coli* gene for D-xylose isomerase [26].

Several hybrid plasmids bearing the gene of interest were isolated which differed in the size of the insert they bore. The one studied in greatest detail had an insert size of 9.7 kb. It contained at least 3 genes involved in D-xylose catabolism, and thus contained elements of the D-xylose operon of *E. coli* (Fig. 8). The genes are

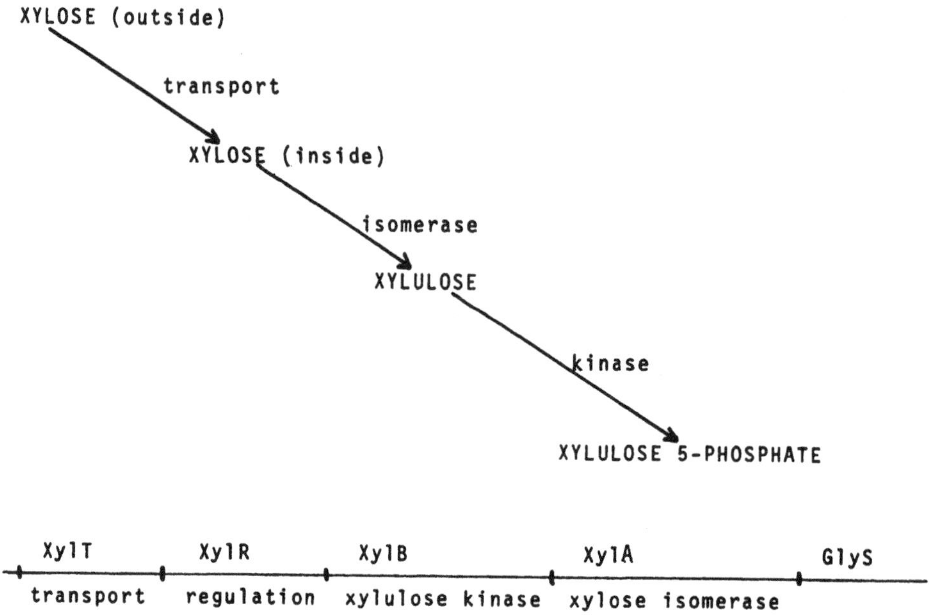

Figure 8. Upper portion: initial steps in the catabolism of D-xylose in *Escherichia coli*[27] and in *Salmonella typhimurium*[28]. Lower portion: order of D-xylose catabolizing genes in the D-xylose operon of *S. typhimurium*[28]

those coding for D-xylulose kinase, D-xylose isomerase and for an associated regulatory element. The cloned segment also bore the structural gene for glycine synthetase. GlyS and Xyl mutations in *E. coli* map at 79 minutes.

Subsequent steps of the strategy require introduction of the D-xylose isomerase gene into an appropriate yeast host, and then obtaining its expression.

6 Overview

Ethanol can be obtained from D-xylose using yeasts. The conversion can be efficient, and at least one organism, *Pachysolen tannophilus*, can carry it out as well with the major hexoses in lignocellulosics. Yeasts can also convert to ethanol, glycerol and mixtures of D-xylose and D-cellobiose, other substrates whose bioconversion in this way is of potential economic importance. A problem which requires attention is that of processing conditions, because of the large effect they have on yield.

Progress in using recombinant DNA methods to construct yeasts that will efficiently convert D-xylose to ethanol has also been made through the cloning of a bacterial gene for D-xylose isomerase. An attractive feature of transforming yeasts to express this gene is that it might be possible to choose a host for which culture conditions could be specified readily, because of existing technology.

7 Acknowledgement

We are indebted to Dr. T. W. Jeffries for critical reading of the manuscript.

8 References

1. Suomalainen, H., Oura, E.: Yeast Nutrition and Solute Uptake in The Yeasts. (Rose, A. H. and Harrison, J. S., eds.) *2*, 9, New York: Academic Press 1971
2. Barnett, J. S.: Adv. Carbohyd. Biochem. *32*, 125 (1968)
3. Schneider, H., Wang, P. Y., Chan, Y. K., Maleszka, R.: Biotechnol. Lett. *3*, 89 (1981)
4. Slininger, P. J., Bothast, R. J., Van Cauwenberge, J. E., Kurzmann, C. P.: Biotech. Bioeng. *24*, 371 (1982)
5. Jeffries, T. W.: Biotechnol. Lett. *3*, 213 (1981)
6. Gong, C. S., McCraken, L. D., Tsao, G. T.: ibid. *3*, 245 (1981)
7. Maleszka, R., Veliky, I. A., Schneider, H.: ibid. *3*, 415 (1981)
8. Veliky, I., Williams, R. E.: ibid. *3*, 275 (1981)
9. Veliky, I., Rose, D.: Can. J. Bot. *51*, 1837 (1973)
10. Forss, K.: The Composion of a Spent Spruce Sulfite Liquor, Diss., Åbo (Finland) 1961. Cited in Rydholm, S. A., Pulping Process, p. 518, New York: Interscience 1965
11. Barnett, J. A., Payne, R. W., Yarrow, D.: A Guide to Identifying and Classifying Yeasts. Cambridge: Cambridge University Press 1979
12. Maleszka, R., Wang, P. Y., Schneider, H.: Enz. Microb. Technol. *4*, 349 (1983)
13. Neirinck, L., Maleszka, R., Schneider, H.: Biotech. Bioeng. Symp. *12* (1982) (in press)
14. Maleszka, R., Schneider, H.: Can. J. Microbiol. *28*, 360 (1982)
15. Desrochers, M., Jurasek, L., Paice, M. G.: Dev. Ind. Micro. *22*, 675 (1981)
16. Maleszka, R., Wang, P. Y., Schneider, H.: Biotechnol. Lett. *4*, 133 (1982)
17. Harrison, J. S., Graham, J. C. J.: Yeasts in Distillery Practice in The Yeasts. (Rose, A. H., Harrison, J. S. eds.), *3*, 283, London and New York: Academic Press 1970
18. Williams, L. S., Foo, E. L., Foo, A. S., Kuhn, I., Hedén, C.-G.: Biotech. Bioeng. Symp. *8*, 115 (1980)
19. Ban, S., Glanser-Soljan, M., Smailagic, M.: Biotech. Bioeng. *21*, 1917 (1979)
20. Maleszka, R., Schneider, H.: Appl. Envir. Micro. *44*, 909 (1982)
21. Karczewska, H.: Compt. Rend. Lab. Carlsberg. *31*, 251 (1959)
22. Karczewska, H.: Swedish Water and Air Pollution Research Institute. IVL Publication B, 463. Stockholm 1978
23. Wang, P, Y., Shopsis, C., Schneider, H.: Biochem. Biophys. Res. Commun. *94*, 248 (1980)
24. Gong, C.-S., Chen, L. F., Flickinger, M. C., Chiang, L. C., Tsao, G. T.: Appl. Env. Microbiol. *41*, 430 (1981)
25. Jeffries, T. W.: Biotech. Bioeng. Symp. *12* (1982) (in press)
26. Maleszka, R., Wang, P. Y., Schneider, H.: Can. J. Biochem. *60*, 144 (1982)
27. David, J. D., Wiesmeyer, H.: Biochim. Biophys. Acta. *201*, 497 (1970)
28. Shamanna, D. K., Sanderson, K. E.: J. Bacteriol. *139*, 64 (1979)

Biology and Physiology of the D-Xylose Fermenting* Yeast *Pachysolen tannophilus*

C. P. Kurtzman
Northern Regional Research Center, Agricultural Research Service, U.S. Dept. of Agriculture, 1815 North University Street, Peoria, Illinois 61604, U.S.A.

The yeast *Pachysolen tannophilus* was found to be capable of converting D-xylose to ethanol. Maximum ethanol yield was 0.34 g per g of pentose consumed. Aerobic conditions were required for cell growth but not for ethanol production. Growth and ethanol formation are optimum under acidic conditions (pH 2.5) at 32C. Among the yeasts, *P. tannophilus* is unique because of its unusual ascosporic stage. Asci are borne on ascophores and form four hatshaped ascospores. The species is homothallic. Comparisons of extracellular polysaccharides suggest *P. tannophilus* to be related to certain tree-inhabiting species of *Hansenula* and *Pichia*.

1 Introduction

Because of their abundant and renewable nature, the use of farm crop residues and other biomass as substrates for the production of liquid fuels and chemical feedstocks has caught the imagination of both scientists and non-scientists. The annual U.S. total for crop residues exceeds 700 million tons. As much as 75% of crop residues may consist of potentially fermentable cellulose and hemicellulose, while the remainder is primarily lignin for which no economically feasible process has been developed (Table 1). Means for converting crop residue fractions into liquid fuel are depicted in Fig. 1. Fungi, because of their enzymatic abilities, have been extensively used for both depolymerization and fermentation of the residues. Cellulases

* Fermentation = Production of ethanol and carbon dioxide under anaerobic conditions

Table 1. Average composition for farm crop residues (given as dry weight %; data from Tsao et al.[1])

Cellulose	34.3
Hemicellulose	39.6
Lignin	19.9
Protein, etc.	4.6
Ash	2.5

are reported from a number of fungi[2,3,4], but those from *Trichoderma reesei* seem to offer the greatest activity. These enzyme complexes degrade cellulose to cellobiose and glucose, which may then be converted to alcohol by yeasts. There are two main obstacles to rapid enzymatic degradation of cellulose:
1) the highly ordered crystalline structure of native fibers, and
2) the presence of lignin around the cellulose fibers (1).

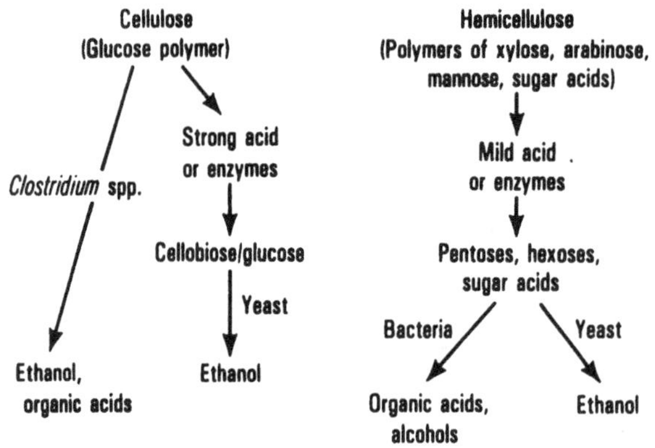

Fig. 1. Potential of cellulose and hemicellulose from crop residues for fermentation to fuel alcohol

Pretreatment of native cellulose with acid renders the molecule more susceptible to enzymatic degradation, but lignin removal is still problematical. However, certain "white rot" fungi, which degrade lignin while leaving most of the cellulose intact, may provide a solution to this problem[5,6,7,8].

The other main fraction of plant residues, hemicellulose, is relatively easily hydrolyzed to its component mono- and oligosaccharides through either mild acid treatment or appropriate enzymes[1]. Pentoses, especially D-xylose, comprise a major portion of hemicellulose. Several species of bacteria ferment pentoses to ethanol, but the process is relatively inefficient because of the concomitant production of short-chain organic acids[9,10]. Direct fermentation of pentoses to alcohol by yeasts has been reported not to occur[11]. However, contrary to this expectation, Schneider et al.[12] and Slininger et al.[13] independently discovered the yeast *Pachysolen tannophilus* to convert D-xylose to ethanol. This finding has now opened the way for more complete utilization of plant residues for liquid fuels and chemical feedstocks.

2 Survey of Yeast Species for D-Xylose Fermentation

Species were selected for D-xylose fermentation tests on the basis of two criteria: 1) ability to oxidatively assimilate D-xylose, and 2) ability to ferment D-glucose. It was felt that species unable to ferment hexoses would be unable to ferment pentoses. Data on D-glucose fermentation and D-xylose assimilation are readily available, since they form part of the taxonomic description of nearly all yeasts. To further reduce the amount of testing, a subset of species was chosen that was comprised primarily of taxa from woody habitats. Although the species surveyed were predominantly from the genera *Hansemula* and *Pichia*, certain species in *Candida* match the selection criteria and would merit examination. Tests were done in Durham tubes using the basal fermentation medium of Wickerham [14]. D-xylose was used at a concentration of 2%.

Of the 54 taxa tested, only *Pachysolen tannophilus* fermented D-xylose (Table 2). Under test conditions, gas formation began 6–8 days after inoculation and reached maximum 7 days later as assessed from carbon dioxide formation. Data from Durham tube tests should be regarded as only semi-quantitative, but the results

Table 2. Yeasts tested for D-xylose fermentation[a, b]

Fermentation of D-xylose	
Pachysolen tannophilus Y-2460, Y-2461, Y-2462, Y-2463, Y-6704	
No fermentation of D-xylose	*P. toletana* YB-4247
Pichia abadiae Y-7499	*P. trehalophila* Y-6781
P. angophorae Y-7118	*P. veronae* Y-7818
P. bovis YB-4184	*P. wickerhamii* Y-2435
P. etchellsii Y-7121	*Hansenula anomala* Y-366
P. fermentans Y-1619	*H. beckii* Y-1482
P. guilliermondii Y-2076	*H. beijerinckii* YB-4312
P. haplophila Y-7860	*H. bimundalis* Y-5343
P. heimii Y-7502	*H. californica* Y-1680
P. lindnerii Y-10948	*H. capsulata* Y-1842
P. membranaefaciens Y-2026	*H. ciferrii* Y-1031
P. methanolica Y-7685	*H. dimennae* YB-3239
P. mucosa YB-1344	*H. fabianii* Y-1871
P. naganishii Y-7654	*H. glucozyma* YB-2185
P. nakazawae var. *nakazawae* Y-7903	*H. henrici* YB-2194
P. nakazawae var. *akitaensis* Y-7904	*H. holstii* Y-2155
P. norvegensis Y-7687	*H. minuta* Y-411
P. onychis Y-7123	*H. mrakii* Y-1364
P. philogaea Y-7813	*H. muscicola* Y-7005
P. pijperi YB-4309	*H. petersonii* Y-3808
P. pinus Y-11528	*H. philodendra* Y-7210
P. rabaulensis Y-11705	*H. polymorpha* Y-5445
P. rhodanensis Y-7854	*H. saturnus* Y-1304
P. sargentensis YB-4139	*H. silvicola* Y-1678
P. scolyti Y-5512	*H. subpelliculosa* Y-1683
P. stipitis Y-7124	*H. sydowiorum* Y-7130
P. strasburgensis Y-2383	*Candida utilis* Y-900, Y-1082, Y-1084

[a] Strain designations are NRRL numbers; [b] Kurtzman, unpublished data

showed an approximately equal fermentation rate for NRRL Y-2460, NRRL Y-2462, NRRL Y-2463 and NRRL Y-6704, whereas NRRL Y-2461 had approximately half that rate.

3 Biology and Physiology of *Pachysolen tannophilus*

3.1 Source of Strains

Pachysolen tannophilus was described by Boidin and Adzet in 1957 [15]. Most strains of this species were isolated from tree extracts used in leather tanning (Table 3). Although *P. tannophilus* reproduces vegetatively by budding in the same manner as many other yeasts, the unusual morphology of its ascosporic state makes it instantly recognizable under the light microscope even at moderate magnifications.

Table 3. History of known strains of *Pachysolen tannophilus*

NRRL No.	Boidin & Adzet No. (I.R.I.C.)	CBS No.	Source and characteristics
Y-2460	145	4044	Extract of *Castanea vesca*, Ludwigshafen, Germany; predominantly diploid; growth mucoid
Y-2461	146		Extract of *Acacia mollissima*, Ludwigshafen, Germany; predominantly haploid; growth mucoid
Y-2462	152		Extract of *Acacia mollissima*; Ludwigshafen, Germany; predominantly haploid; growth mucoid
Y-2463	153		Shoe leather in association with *Penicillium* sp., Strasbourg, France; predominantly haploid; growth mucoid
Y-6704	164	4045	Extract of *Acacia mollissima*; Ludwigshafen, Germany; predominantly haploid; growth mat, not mucoid. This strain was originally described as *Pachysolen pelliculatus*.

3.2 Vegetative Reproduction

Vegetative growth is comprised mainly of budding yeast cells that are spheroidal to ellipsoidal. After 3 days at 25 C, cells on malt extract agar measure 1.5–5.0 × 2.0 to 7.0 μm and usually have one or two buds. Pseudohyphae are usually present and may be undifferentiated or highly branched.

3.3 Sexual Reproduction

Ascus formation is unique. Initially, a vegetative cell produces a stout tube which may be quite short or up to 60 μm in length and either straight or curved. The tip of the tube enlarges to form the ascus; consequently, the tube may be regarded as an ascophore. Asci contain up to four hemispheroidal ascospores that have a narrow ledge at the base. The ascus wall deliqueses, releasing the spores. Once this happens,

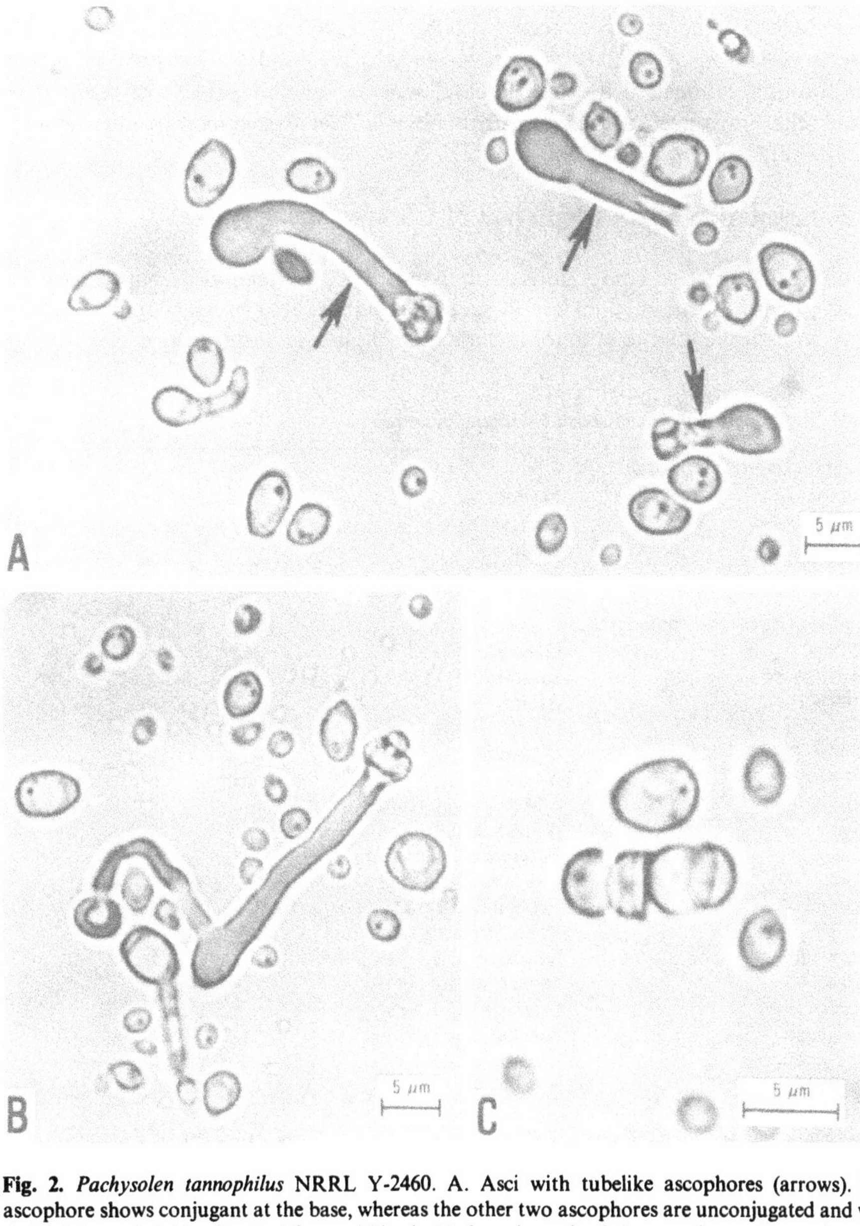

Fig. 2. *Pachysolen tannophilus* NRRL Y-2460. A. Asci with tubelike ascophores (arrows). Left ascophore shows conjugant at the base, whereas the other two ascophores are unconjugated and were derived from diploid cells. Asci form within the V-shaped notch of the ascophores. Vegetative cells, some with one or two buds, are also present. B. Ascophore and conjugating cell connected by a relatively long conjugation tube. C. Linear cluster of 4 hat-shaped ascospores. Ascospores become swollen upon release from the ascus

it is seen that the ascus has formed within a V-shaped notch at the end of the ascophore. Ascophore walls appear greatly thickened and become quite refractile. Ascosporogenous cultures may be immediately identified to species under the light microscope because of the uniqueness of the ascus (Fig. 2).

Asci may be conjugated or unconjugated, and this is strain dependent to some extent. Most of the asci produced by NRRL Y-2460 are unconjugated, but the majority produced by NRRL Y-2461, Y-2462, Y-2463 and Y-6704 are conjugated. Cells forming conjugated asci are haploid whereas unconjugated asci result from diploid cells. Single ascospore isolates from NRRL Y-2460 gave sporogenous colonies, indicating the species to be homothallic [16].

3.4 Fermentation and Assimilation of Carbon Compounds

The number of carbon compounds assimilated by *P. tannophilus* is limited to a few hexoses, pentoses, organic acids and sugar alcohols (Table 4). Carbon compounds known to be fermented to ethanol include only D-glucose and D-xylose.

Table 4. Physiological characteristics of *Pachysolen tannophilus*[a]

Fermentation:	
D-Glucose +	Maltose –
D-Galactose –	Lactose –
D-Xylose +	Raffinose –
Sucrose –	
Assimilation of carbon compounds:	
D-Glucose +	Glycerol +
D-Galactose +	Erythritol –
L-Sorbose –	Ribitol +
Sucrose –	Galactitol –
Maltose –	D-Mannitol +
Cellobiose +	D-Glucitol +
Trehalose –	α-Methyl-D-glucoside –
Lactose –	D-Glucosamine w or –
Melibiose –	Potassium D-gluconate –
Raffinose –	Calcium 2-keto-D-gluconate –
Melezitose –	Potassium 5-keto-D-gluconate –
Inulin –	Potassium acid saccharate –
Soluble starch –	Salicin + or –
D-Xylose +	Pyruvic acid + or –
L-Arabinose +	DL-Lactic acid –
D-Arabinose –	Succinic acid +
D-Ribose + or –	Citric acid –
L-Rhamnose –	Inositol –
Ethanol +	

Assimilation of potassium nitrate: positive.
Growth in vitamin-free medium: negative.
Growth in 10% sodium chloride plus 5% glucose in yeast nitrogen base: moderate.
Growth at 37 C: positive.
Production of esters: positive, a sweet, pleasant smell.
Guanine + cytosine content of the nuclear DNA: 43.0 mol% [28]

[a] – = negative; + = positive; w = weakly positive

3.5 Extracellular Polysaccharide

The mucoid nature of cultures of *P. tannophilus* demonstrates the presence of an extracellular polysaccharide, which, depending upon the amount of orthophosphate in the medium, was shown to be comprised of either *O*-phosphonomannans or α-D-mannans [17,18]. Methylation analysis and acetolysis studies [19] revealed the mannans to be highly branched with (1→2)-linked side-chains attached primarily through (1→3)-linkages to (1→6)-linked backbone D-mannosyl residues. Some 2,3,4-tri-*O*-methyl-D-mannose in addition to 3,4-di-*O*-methyl-D-mannose, were also present in the hydrolyzate of the permethylated mannan. Slodki et al. [18] determined the yield of extracellular polysaccharide to be 3 g l^{-1} in a medium containing 5% glucose. Consequently, there is a need to account for polysaccharide when measuring process efficiency. Indeed, D-mannan production accounts for a significant portion of the D-xylose not converted to alcohol (K. L. Smiley and M. E. Slodki, unpublished observations).

3.6 Taxonomy and Phylogeny

In Boidin and Adzet's [15] description of the genus *Pachysolen*, two species were defined, *P. tannophilus* and *P. pelliculatus*. The main differences in the descriptions of the two species were mat growth on solid media; presence of a thin, dry pellicle on liquid media; formation of more pseudohyphae and a weaker growth of glucose for *P. pelliculatus*. Wickerham [16] reasoned that *P. pelliculatus* represented a mat form of *P. tannophilus*. To demonstrate this, he selected for the mat form in *P. tannophilus* by allowing an amoeba to feed on the culture. The amoeba preferentially ingested mucoid cells allowing the development and establishment of a mat colony type of *P. tannophilus* indistinguishable from *P. pelliculatus*. From this he concluded that the two taxa were conspecific and that *P. pelliculatus* must be regarded as a synonym of *P. tannophilus*. In 1973, on the basis of numerical analysis, Campbell [20] transferred *P. tannophilus* to the genus *Hansenula*, but that work has not been accepted by other taxonomists because they felt that morphological differences in the ascosporic states of the two genera were too great.

The relationship of *Pachysolen* with other ascomycetous yeast genera is uncertain. A comparison of mannans and phosphomannans from several taxa suggests *P. tannophilus* to have an evolutionary kinship with certain tree-inhabiting species in *Hansenula* and *Pichia* [21]. Resolution of this question will need to await molecular studies such as comparisons of DNA/RNA complementarity.

4 Conditions for D-Xylose Fermentation by *Pachysolen tannophilus*

Conditions governing the production of ethanol from D-xylose by *P. tannophilus* were dealt with in detail by Slininger et al. [13] and only key points will be summarized here. The influence of temperature on ethanol formation was tested over a range of 15–40 C. Maximum yield occurred at 32 C (Table 5). The maximum temperature

Table 5. Effect of temperature on growth, ethanol production, and D-xylose consumption by *P. tannophilus* (data from Slininger et al. [13])

Temperature C	Maximum specific growth rate h^{-1}	Maximum specific ethanol production rate $g\,g^{-1}\,h^{-1}$	Maximum specific xylose consumption rate $g\,g^{-1}\,h^{-1}$	Ethanol yield $g\,g^{-1}$
15	0.09	0.05	0.12	0.26
20	0.14	0.06	0.26	0.27
25	0.16	0.07	0.32	0.27
28	0.19	0.11	0.44	0.28
32	0.24	0.12	0.49	0.34
40	0.06	0	0	—

Table 6. Effect of pH on growth, ethanol production, and D-xylose consumption by *P. tannophilus* (data from Slininger et al. [13])

pH	Maximum specific growth rate h^{-1}	Maximum specific ethanol production rate $g\,g^{-1}\,h^{-1}$	Maximum specific xylose consumption rate $g\,g^{-1}\,h^{-1}$	Ethanol yield $g\,g^{-1}$
2.5	0.18	0.13	0.43	0.20
4.5	0.20	0.07	0.32	0.22
6.5	0.09	0.03	0.39	0.08
7.5	0.09	0	0	—

Table 7. Effect of initial substrate concentration on growth, ethanol production, and D-xylose consumption by *P. tannophilus* (data from Slininger et al. [13])

Initial xylose concentration $g\,l^{-1}$	Initial specific growth rate h^{-1}	Maximum specific ethanol production rate $g\,g^{-1}\,h^{-1}$	Maximum specific xylose consumption rate $g\,g^{-1}\,h^{-1}$	Ethanol yield $g\,g^{-1}$
55	0.22	0.09	0.3	0.29
115	0.19	0.09	0.4	0.24
158	0.13	0.08	0.4	0.21
210	0.09	0.06	0.4	—
255	0.07	0.05	0.4	—

for growth is between 37 and 40 C. A comparison of fermentation efficiency at pH values between 2.5 and 7.5 showed maximum specific ethanol production to occur at pH 2.5 (Table 6). Such low acidity would suppress most bacterial contaminants that might be found under commercial process conditions. Fermentation activity with initial D-xylose concentrations of 55 to 255 g l^{-1} was maximal at the lowest concentration tested (Table 7).

Data presented by Schneider et al. [12] indicated that oxygen was required for the conversion of D-xylose to ethanol by *P. tannophilus*. Slininger et al. [13] also addressed this issue. Their data showed essentially no cell growth in the absence of oxygen and, therefore, no fermentation when the inoculum level was low. However, when the medium was inoculated with a large cell mass, ethanol production occurred under anaerobic conditions despite the absence of cell multiplication. This is demonstrated in Fig. 3.

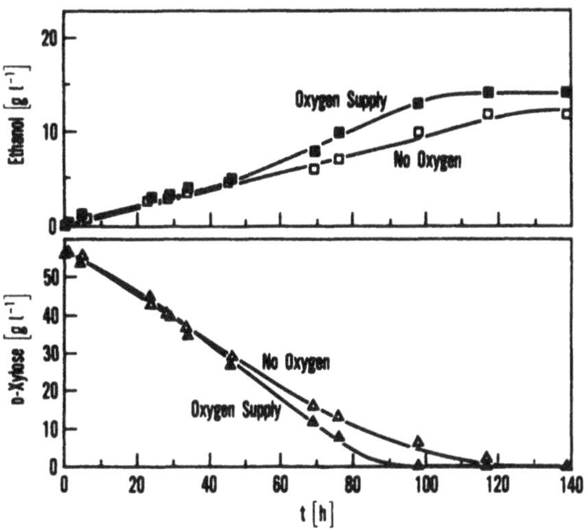

Fig. 3. Influence of oxygen on ethanol production and D-xylose consumption in shake cultures inoculated with a large cell population of *P. tannophilus*. After Slininger et al. [13]

An investigation of D-xylose fermentation by *P. tannophilus* on a continuous basis was reported by Slininger et al. [22]. Yeast cells were entrapped in calcium alginate beads and supplied continuously with fresh medium under anaerobic conditions. Rate of ethanol production was dependent on substrate level with the optimum occurring when the D-xylose concentration was between 28 and 35 g l^{-1}. Fermentation efficiency was highly dependent upon D-xylose concentration. The yield was 69% of that estimated to be theoretically possible when the D-xylose concentration was near zero as contrasted to 42% when it was in the range supporting the optimum rate of ethanol production. Under favorable conditions, the immobilized cells retained at least 50% of their initial productivity after 26 d of operation. However, bead size had a marked influence on availability of substrate to the immobilized cells. The 4 mm diameter beads were shown from calculations based on the effectiveness factor concept to be only 40% efficient thus demonstrating the need for a smaller diameter.

Detroy et al. [23] examined the ability of *P. tannophilus* to ferment the D-xylose from crude wheat straw hydrolyzates. No fermentation occurred in preparation obtained by either autohydrolysis or acid hydrolysis. Failure to ferment was attributed to toxic concentrations of furfural and lignin derivatives in the hydrolyzates. Furfural at concentrations of 0.25–0.30% was lethal to *P. tannophilus*. Extraction of the hydrolyzates with either ethanol or diethyl ether successfully removed the toxic compounds and permitted fermentation. Ether was the more efficient extractant since ethanol removed as much as 25% of the pentosans in the hydrolyzates.

5 Concluding Remarks

Pachysolen tannophilus is the first yeast known to be capable of fermenting pentoses to ethanol. This finding is an important step toward efficiently converting the pentoses of plant residues to liquid fuels. Because *P. tannophilus* tolerates little more than 3–6% ethanol, strain improvements seem necessary. Further, Detroy et al. [23] point out that crude straw hydrolyzates are inhibitory to *P. tannophilus* and require some purification of degraded residues before fermentation can occur.

Because *P. tannophilus* is homothallic, strain improvement through conventional genetic techniques may be difficult to realize. Further, it is not yet determined what genetic diversity the five known strains possess. It is clear, however, that genetic modification of some sort should be attempted to increase the efficiency of pentose fermentation. Once the biochemical pathway for D-xylose fermentation is elucidated, the genes controlling this process might be transferred through molecular cloning techniques to yeasts more tolerant of crude substrates and high ethanol concentrations. One species in particular, *Candida utilis*, is well adapted to sulfite waste liquor from paper pulping and is commonly used to oxidatively assimilate D-xylose from this processing waste [24]. Another potential recipient of D-xylose fermentation genes is *Candida tropicalis* which commonly contaminates *C. utilis* growth in sulfite waste liquor. If molecular cloning proves difficult, fusion of protoplasts from *P. tannophilus* and another yeast may be practical.

The discovery of D-xylose fermentation by *P. tannophilus* suggests this property will be present in still other fungi. The recent reports of D-xylose fermentation by *Candida tropicalis* [25], *Candida* sp. [26] and *Fusarium* spp. [27] bear out this expectation.

6 References

1. Tsao, G. T. et al.: Fermentation substrates from cellulosic materials: Production of fermentable sugars from cellulosic materials, in: Annu. Rep. Ferment. Processes, Vol. 2, (Perlman, D., Tsao, G. T. eds.), p. 1, Academic Press 1978
2. Emert, G. H. et al.: Adv. Chem. Ser. *136*, 79 (1974)
3. Eberhart, B. M., Beck, R. S., Goolsby, K. M.: J. Bacteriol. *130*, 181 (1977)
4. Eriksson, K. E., Hamp, S. G.: Eur. J. Biochem. *90*, 183 (1978)
5. Kirk, T. K.: Annu. Rev. Phytopathol. *9*, 185 (1971)
6. Crawford, D. L., Crawford, R. I.: Appl. Environ. Microbiol. *31*, 714 (1976)
7. Kaneshiro, T.: Dev. Ind. Microbiol. *18*, 591 (1977)
8. Wicklow, D. T., Detroy, R. W., Jessee, B. A.: Appl. Environ. Microbiol. *40*, 169 (1980)
9. Rosenberg, S. L.: Enzyme Microbiol. Technol. *2*, 185 (1980)
10. Wood, W. A.: Basic biology of microbial fermentation, in: Trends in the biology of fermentations for fuel and chemicals. (Hollaender, A., Rabson, R., Rogers, P., San Pietro, A., Valentine, R., Wolfe, R. eds.), p. 3, Plenum Press 1981
11. Barnett, J. A.: Adv. Carbohydr. Chem. Biochem. *32*, 125 (1976)
12. Schneider, H., et al.: Biotechnol. Lett. *3*, 89 (1981)
13. Slininger, P. J., et al.: Biotech. Bioeng. *24*, 371 (1982)
14. Wickerham, L. J.: U.S. Dep. Agric., Tech. Bull. No. 1029 (1951)
15. Boidin, J., Adzet, J.-M.: Bull. Soc. Mycol. France *73*, 331 (1957)
16. Wickerham, L. J.: *Pachysolen* Boidin et Adzet, in: The yeasts, a taxonomic study. (Lodder, J. ed.), p. 448, North-Holland 1970
17. Slodki, M. E., Wickerham, L. J., Cadmus, M. C.: J. Bacteriol. *82*, 269 (1961)

18. Slodki, M. E., Ward, R. M., Cadmus, M. C.: Dev. Ind. Microbiol. *13*, 428 (1972)
19. Seymour, F. R., et al.: Carbohydr. Res. *48*, 225 (1976)
20. Campbell, I.: J. Gen. Microbiol. *77*, 427 (1973)
21. Wickerham, L. J., Burton, K. A.: J. Bacteriol. *82*, 265 (1961)
22. Slininger, P. J., et al.: Biotech. Bioeng. *24*, 371 (1982)
23. Detroy, R. W., Cunningham, R. L., Herman, A. I.: Biotech. Bioeng. Symp. No. *12*, 81 (1982)
24. Peppler, H. J.: Food yeasts, in: The yeasts, Vol. 3, Yeast technology. (Rose, A. H., Harrison, J. S. eds.), p. 421, Academic Press 1970
25. Jeffries, T. W.: Biotechnol. Lett. *3*, 213 (1981)
26. Gong, C.-S., McCracken, L. D., Tsao, G. T.: ibid. *3*, 245 (1981)
27. Suihko, M.-L., Enari, T.-M.: ibid. *3*, 723 (1981)
28. Nakase, T., Komagata, K.: J. Gen. Appl. Microbiol. *14*, 345 (1968)

Bioconversion of Pentoses to 2,3-Butanediol by *Klebsiella pneumoniae*

Norman B. Jansen and George T. Tsao
Laboratory of Renewable Resources Engineering, Purdue University, West Lafayette, IN 47907, U.S.A.

The facultative anaerobe *Klebsiella pneumoniae* converts a wide variety of sugars to 2,3-butanediol. The theoretical maximum yield of butanediol from sugar is 0.50 kg per kg. All of the sugars commonly found in hemicellulose and cellulose hydrolysates can be converted to butanediol, including glucose, xylose, arabinose, mannose, galactose, and cellobiose. In contrast with most yeast, the inducibility of the key enzyme, xylose isomerase, facilitates rapid metabolism of xylose. In this report the metabolic pathways leading from pentoses to 2,3-butanediol are reviewed. In addition, some of the key factors affecting butanediol production are discussed. These variables include temperature, pH, sugar concentration, oxygen supply, and water activity.

1 Introduction

Due to a projected long-term shortage of crude oil, the possibility of producing chemical feedstocks and liquid fuels from annually renewable resources is receiving considerable interest. Common sources of renewable resources include corn stover, sugarcane bagasse, wood chips, municipal wastes, and animal feedlot wastes [1,2]. Large quantities of these materials are available for utilization in processes leading to the production of more valuable substances [3-7]. The biomass in these materials is composed mainly of cellulose, hemicellulose, and lignin. Many of these cellulosic materials contain nearly as much hemicellulose as cellulose [8,9].

Some of the older processes for the conversion of biomass into useful products considered only the cellulose fraction, and the hemicellulose was lost [10]. In order to make biomass conversion economically feasible, it is essential that the hemicellulose fraction also be efficiently converted into useful products [11]. While biomass derived chemicals may never replace petrochemical feedstocks, it is possible to visualize a scheme whereby fermentation products make an important contribution to the supply of feedstocks for the chemical industry [12–14].

Hemicellulose is a highly branched, heterogeneous polymer composed mainly of pentoses. It is easily hydrolyzed to D-xylose, L-arabinose, D-mannose, and D-galactose. D-Xylose generally accounts for over 60% of the monomeric sugars present in hemicellulose [15]. Although most yeast cannot utilize xylose anaerobically, many bacteria readily convert xylose to a variety of products in the absence of oxygen [16].

One example of an interesting use for xylose is the production of 2,3-butanediol by *Klebsiella pneumoniae*. Butanediol may be valuable as a chemical feedstock or as a liquid fuel. Since the value of a reactive compound as a chemical feedstock generally exceeds its fuel value, 2,3-butanediol may be more valuable as a chemical feedstock. The dehydration of 2,3-butanediol yields the common industrial solvent methyl ethyl ketone [17]. Further dehydration yields 1,3-butadiene, which is the monomeric subunit in synthetic rubber [18]. Dimerization of butadiene by the Diels-Alder reaction produces styrene, an important aromatic intermediate [14]. Styrene and butadiene are both important monomers in the polymers industry. It is also possible to prepare a variety of other chemicals from 2,3-butanediol [19].

One of the advantages of 2,3-butanediol production is that *K. pneumoniae* is easy to cultivate. It grows rapidly in a simple medium [20] and it is able to metabolize a wide variety of sugars [21]. If genetic improvement of *K. pneumoniae* becomes desirable, its similarity to *E. coli* could prove to be a fortuitous advantage. Perhaps the most important attribute of *K. pneumoniae* related to biomass conversion is its ability to convert all of the major sugars present in hemicellulose and cellulose hydrolysates into 2,3-butanediol [22].

2 Biochemical Pathways Leading from Pentoses to 2,3-Butanediol

2.1 Conversion of Pentoses to D-Xylulose-3-Phosphate

Klebsiella pneumoniae grows on 7 of the 8 aldopentoses, even though several of these compounds rarely occur in nature [23]. Figure 1 depicts the pathways of pentose and pentitol metabolism by this organism as determined by Mortlock [23]. The aldopentoses are isomerized to one of the four ketopentoses by a pentose isomerase. This ketopentose is phosphorylated by a pentulokinase and then epimerized to D-xylulose-5-phosphate. The pentose isomerases and pentulokinases required for the metabolism of D-xylose, L-arabinose, and D-ribose are induced rapidly after exposure of the cells to the pentose [24]. However, when cells are exposed to L-xylose, D-arabinose, or

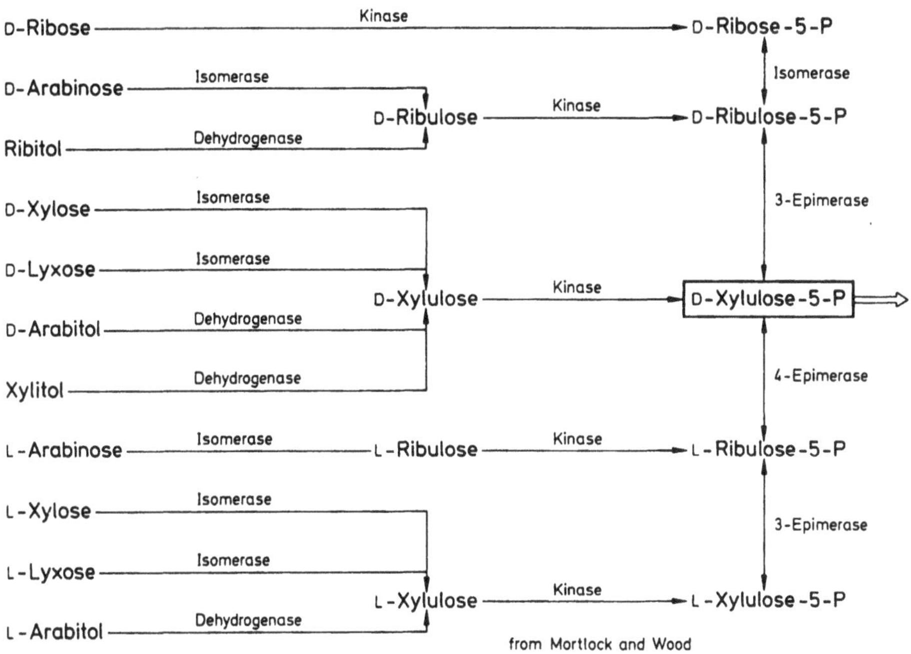

Fig. 1. Conversion of pentoses and pentitols to D-xylulose-5-phosphate [23)]

D-xylose, there is a long lag time before the enzymes required for assimilation appear. This lag may indicate that a mutation is required before growth on these sugars is possible [24)]. It is thought that the result of this mutation is that the inducible isomerizing enzyme becomes constitutive [25)].

It is interesting to note that many yeasts are unable to metabolize D-xylose [26)]. This difficulty may be due to the lack of the enzyme, xylose isomerase. However, the inducibility of xylose isomerase in *K. pneumoniae* facilitates rapid growth on D-xylose and other pentoses.

2.2 Conversion of D-Xylulose-5-Phosphate to D-Glyceraldehyde-3-Phosphate

As shown by Fig. 1, pentoses are all metabolized through the common intermediate D-xylulose-5-phosphate. In many microorganisms, xylulose-5-P is cleaved to form one three-carbon unit and one two-carbon unit [27)]. This usually results in a mixture of products. On the other hand, *K. pneumoniae* completely converts xylulose-5-P to a triose via the transketolase and transaldolase enzymes of the pentose phosphate pathway. The metabolic scheme illustrated in Fig. 2 was deduced by Neish, et al., from experiments using labelled pentoses [28,29)]. The net result of the pentose phosphate pathway is the conversion of 3 moles of pentose to 5 moles of D-glyceraldehyde-3-phosphate with the consumption of 5 moles of ATP. In terms of yield, this conversion is equivalent to the glycolysis of D-glucose to D-glyceraldehyde-3-P. Thus, for *K. pneumoniae*, product yields from pentoses should be the same as the yields from glucose. Other than differences in the uptake and transport of xylose as compared to glucose, the glycolysis of glucose is energetically equivalent to the metabolism of xylose. Each pathway requires one ATP molecule per molecule of glyceraldehyde-3-P formed.

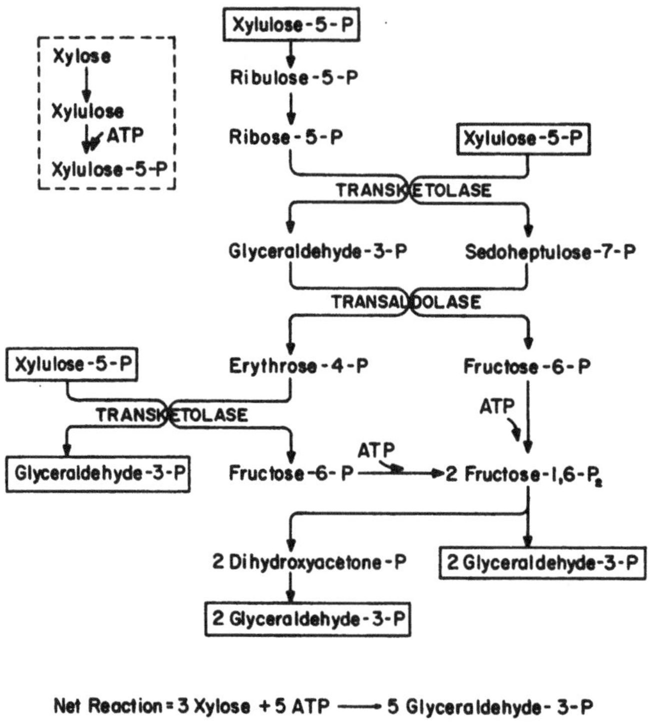

Fig. 2. Pentose phosphate pathway

2.3 Conversion of D-Glyceraldehyde-3-Phosphate to 2,3-Butanediol

The glyceraldehyde-3-P formed from pentoses or hexoses is converted to pyruvate by the constitutive enzymes of the lower glycolytic pathway. *K. pneumoniae* is a facultative anaerobe which means it can obtain all the biological energy it needs for growth either aerobically or anaerobically. Pyruvate is the point at which the catabolic reactions diverge to two different but parallel energy-producing pathways. Most of the pyruvate is either oxidized to carbon dioxide by the tricarboxylic acid cycle, or is converted to various products by anaerobic "fermentative" pathways.

The predominant product of the oxygen-limited growth of *K. pneumoniae* is 2.3-butanediol. The reactions leading from pyruvate to butanediol were first deduced by

Juni[30]. Figure 3 shows this reaction sequence. The first step is the condensation of two moles of pyruvate to form α-acetolactate. This reaction is catalyzed by the "pH 6 acetolactate-forming enzyme"[31,32]. The second step, catalyzed by acetolactate decarboxylase, is the conversion of α-acetolactate to acetoin (acetyl methyl carbinol)[33]. The final step is the reduction of acetoin to 2,3-butanediol by acetoin

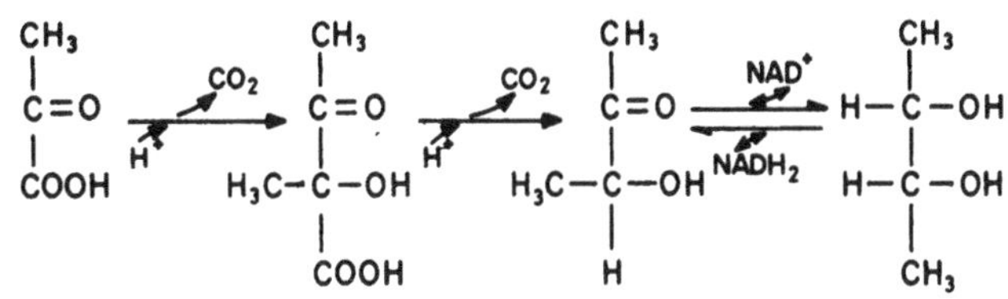

Fig. 3. Conversion of pyruvate to 2,3-butanediol

reductase[34]. This reaction helps maintain the $NAD^+/NADH_2$ balance inside the cell by oxidizing some of the excess reducing equivalents which are produced by the anaerobic reactions. Purification of the enzyme acetoin reductase has shown that it is actually two different stereospecific enzymes. One reduces L-acetoin to L-butanediol and the other reduces D-acetoin to *meso*-butanediol[35]. *K. pneumoniae* produces about 5%–14% L-butanediol and the remaining 86%–95% is *meso*. Both isomers are potentially valuable chemical feedstocks[19].

2.4 Product Yields

The net equation for the reactions leading to 2,3-butanediol is

$$\text{Xylose} \rightarrow \frac{5}{3}\,CO_2 + \frac{5}{6}\,NADH_2 + \frac{5}{3}\,ATP + \frac{5}{6}\,\text{Butanediol} \tag{1}$$

The equation describing the conversion of glucose to 2,3-butanediol is

$$\text{Glucose} \rightarrow 2\,CO_2 + NADH_2 + 2\,ATP + \text{Butanediol} \tag{2}$$

Of the total carbon contained in the sugar, $^2/_3$ goes to 2,3-butanediol, and the other $^1/_3$ is lost as carbon dioxide. On a mass basis, the yield of butanediol from both xylose and glucose is 50%. The molar yield of butanediol from pentoses is 0.83, while from hexoses it is 1.0. These values are the theoretical maximum yields. They represent the maximum value the yield can be if no sugar is assimilated to cell mass or oxidized by the tricarboxylic acid cycle. The actual yield cannot exceed these values.

3 Factors Affecting 2,3-Butanediol Production

3.1 Introduction

The metabolic pathways discussed in the previous section show that xylose and glucose metabolism are very similar. The only differences are the reactions leading to glyceraldehyde-3-phosphate and possibly sugar uptake. Recent experiments with *K. pneumoniae* have demonstrated that the products of xylose metabolism are the same as for glucose metabolism [36)]. Due to a shortage of data concerning butanediol production from pentoses, and because of the similarities between hexose and pentose metabolism, the research reviewed in this section includes experiments utilizing hexoses as well as pentoses as the carbon source.

3.2 Temperature

In continuous culture experiments with *K. pneumoniae*, Pirt measured a maximum sucrose uptake rate at 37 C. In the same experiments, maximum butanediol production occurred between 35 and 37 C [37)].

Topiwala and Sinclair also studied the continuous culture growth of *K. pneumoniae* growing on glucose as a function of temperature. They modelled kinetics by assuming a Monod growth equation along with a maintenance term to correct for endogenous metabolism [38)]. Substrate utilization for endogenous metabolism increased steadily as the temperature increased from 25 to 40 C. Meanwhile the maximum specific growth rate occurred at 38 C [38)].

For batch growth, Esener et al. used an Arrhenius type model to describe the dependence of the exponential growth rate on temperature. Assuming an Arrhenius growth term and an Arrhenius term for enzyme inactivation at higher temperatures, the maximum specific growth rate on glycerol is expected at 37 C [39)].

It is interesting to note that while the optimum temperature for growth and product formation appears to be approximately 37 C, the maximum butanediol concentration ever obtained (99 g l^{-1}) occurred at 30 C [40)].

3.3 pH

Figure 4 illustrates the effect of pH on product yields for the anaerobic conversion of glucose by *K. pneumoniae* [41)]. In this experiment anaerobic conditions were maintained by nitrogen sparge, and pH control was achieved with ammonium hydroxide. Carbon balances for each experiment indicate the fraction of metabolized glucose that appears in each product. The predominant product is 2,3-butanediol. Ethanol is also a major product of anaerobic conversion, accounting for nearly 20% of the metabolized glucose carbon at all pH levels. The ratio of butanediol to acetoin varies from near 25 between pH 5.2 and pH 6.0, to near zero at pH 7.6. The yield of butanediol reaches a maximum in the range pH 5.2–5.6, but falls to near zero above pH 7 (Fig. 4). A rising formic acid yield and falling carbon dioxide yield above pH 7 suggests that the cell is able to maintain its $NAD^+/NADH_2$ balance by reducing carbon dioxide to formic acid when it cannot produce 2,3-butanediol. Hydrogen is also a product of anaerobic turnover at all pH levels[41]. Acetic acid production is minimum in the range pH 5.2–5.6, but increases rapidly above pH 6. Lactic acid production also increases tenfold between pH 5.6 and pH 7[41].

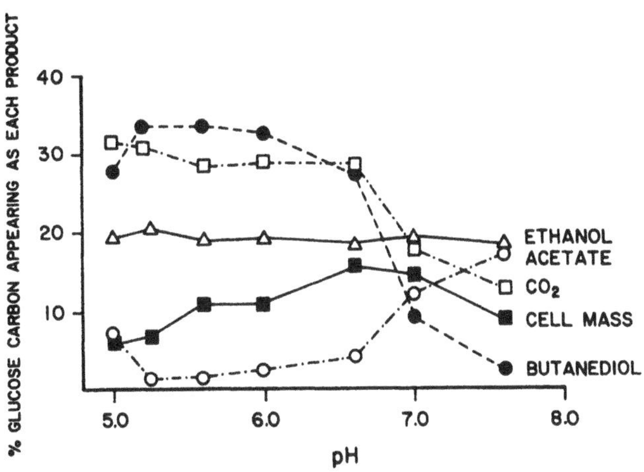

Fig. 4. Product yields versus pH for the anaerobic dissimilation of glucose[41]

For "partially aerobic" growth, the optimum pH for butanediol production from sucrose occurred between pH 5.0 and pH 6.0[37]. Lactic acid production also increased above pH 6 in this continuous culture study. The specific uptake rate of sucrose was 10% higher at pH 5.5 than it was at either pH 5.0 or pH 6.0[37].

When xylose was the sole carbon and energy source, the optimum butanediol yield also occurred between pH 5.0 and pH 6.0, and above pH 6.5 the yield declined sharply[36]. For these oxygen limited experiments the product yields were similar to the anaerobic yields depicted by Fig. 4 except that the ethanol yields were lower while the butanediol yields were higher.

3.4 Sugar Concentration

One of the earliest studies concerning 2,3-butanediol production demonstrated that sucrose concentration affects both product yields and reaction rates[42]. Butanediol yield and production rate were both maximum at a sucrose concentration of 80 g l^{-1}. At higher sugar concentrations, the yield and rate both decreased[42].

The optimum sugar concentration often depended on the particular raw material used as the carbon source. Long has suggested that as the sugar concentration in the raw material is increased, the level of accompanying toxic material also increases, resulting in poor substrate utilization[43]. Using a defined medium with 100 g l^{-1}

glucose, all the sugar was rapidly consumed[40]. However when an acid-hydrolyzed wheat-mash medium was employed, the butanediol yield fell and carbohydrate utilization became incomplete at sugar concentrations over 90 g l^{-1} [40].

As a means of increasing the amount of sugar utilized without increasing the sugar concentration above an optimal level, Olson and Johnson employed a "slow feed" bioreactor[40]. By slowly adding a concentrated glucose solution, they maintained the sugar level in the fermentor near 30 g l^{-1}. Using this system they reported utilizing 265 g l^{-1} glucose.

When xylose was the substrate, maximum cell growth occurred at a sugar concentration of 20 g l^{-1} (Fig. 5). These batch experiments were run at pH 5.2 in a minimal salts synthetic medium[36]. The maximum specific growth rate of 1.05 h^{-1} corresponds to a doubling time of 40 minutes. The apparent substrate inhibition at xylose concentrations over 20 g l^{-1} may be explained by a decreasing water activity[44].

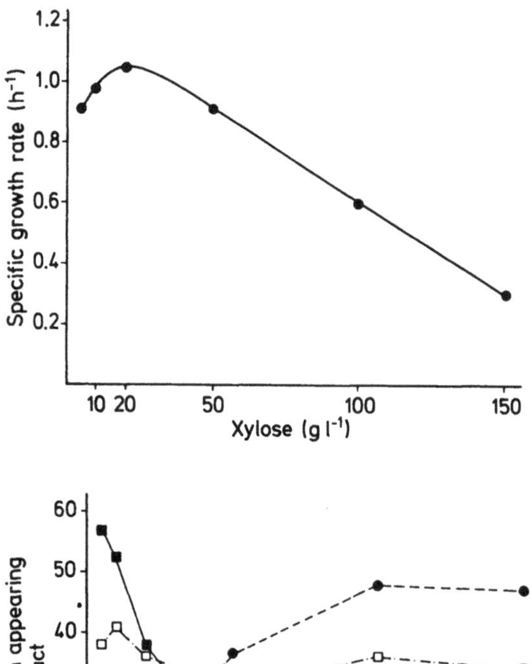

Fig. 5. Maximum specific growth rate versus initial xylose concentration at pH 5.2[36]

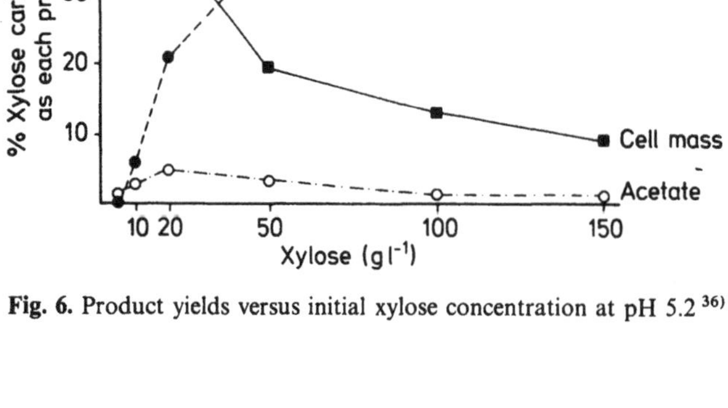

Fig. 6. Product yields versus initial xylose concentration at pH 5.2[36]

Figure 6 portrays the final product yields as a function of the initial xylose concentration. In this plot, 2,3-butanediol also includes the oxidized form, acetoin, which accounted for less than 10% of the total. As illustrated by Fig. 6, at higher initial xylose levels the butanediol yield increases while the cell yield decreases [36]. However, this behavior may be only an indirect result of xylose concentration.

The six experiments depicted by Fig. 6 were run under conditions of constant aeration and agitation. The oxygen supply rate was 27 millimoles per liter per hour. At low xylose levels, less total cell mass is produced, so the oxygen supply *per cell* is greater. This larger specific oxygen supply may allow more substrate to be metabolized through the "preferred" respiratory pathways. As a result, little butanediol is produced at low xylose levels [36]. On the other hand, at higher sugar concentrations, higher cell densities are developed, so the oxygen supply per cell is smaller. This may cause a larger proportion of the cell's energy requirement to be met by the production of 2,3-butanediol via the anaerobic pathways. Therefore, the butanediol yield is higher at the higher initial xylose levels due to a lower average specific oxygen uptake rate [36].

3.5 Oxygen Supply Rate

Although 2,3-butanediol is a product of anaerobic metabolism, aeration has been shown to increase its production by *K. pneumoniae* [19]. The mechanism of this enhancement was originally thought to have been due to the removal of carbon dioxide from the medium [43]. More recently, Pirt has demonstrated that aeration increases butanediol productivity by increasing the cell concentration [45].

In studies with a multistage tower fermentor, Paca demonstrated that when the oxygen transfer coefficient increased, the cell yield from glucose increased while the respiratory quotient decreased toward 1.0 [46]. This suggests a suppression of "fermentative" metabolism by oxygen, It appears that the cell yield and the product yield are both functions of the oxygen supply rate.

With xylose as the substrate, Fig. 7 illustrates the effect of the oxygen supply rate on the final product yields of four batch experiments. These experiments were

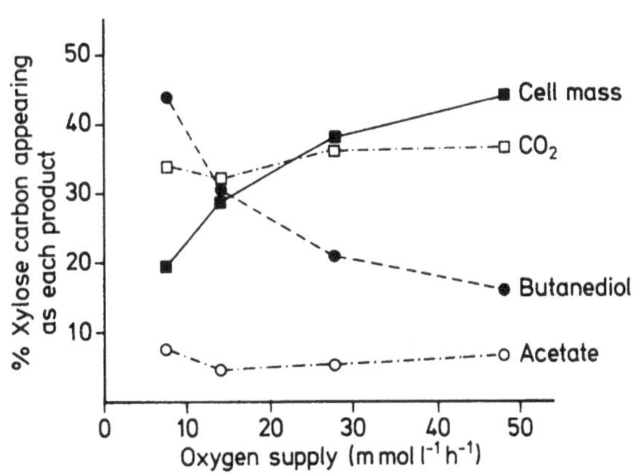

Fig. 7. Product yields versus oxygen supply rate at pH 5.2 with 20 g l^{-1} xylose [36]

run at pH 5.2 with an initial xylose concentration of 20 g l^{-1}. At the maximum oxygen transfer rate, the butanediol yield is lowest and the cell yield is highest (Fig. 7). In all four of these experiments, the dissolved oxygen level was near zero during most of the experiment, and the bacteria obtain energy by both respiration to carbon dioxide and "fermentation" to various products [36]. At the higher levels of oxygen supply, more oxygen is transferred to the culture, and a larger proportion of the cell's energy requirement is met by respiration. This results in a lower butanediol yield at higher oxygen supply rates. On the other hand, at the lowest oxygen transfer rate, less respiration is possible and hence, a larger proportion of the cell's energy requirement must be met by the "fermentative" production of 2,3-butanediol [36].

By decreasing the oxygen transfer rate toward zero, the butanediol yield may be increased toward the theoretical maximum. However, the total reaction rate decreases significantly due to a diminishing cell yield. Yu and Saddler have reported that *K. pneumoniae* will metabolize glucose anaerobically, but that at least some air is required for xylose metabolism [22]. The oxygen limited growth rate of *K. pneumoniae* on xylose appears to be directly proportional to the oxygen transfer rate [36].

The yields depicted by Fig. 7 demonstrate that the oxygen supply influences the balance between respiratory metabolism and "fermentative" metabolism, both of which occur simultaneously during oxygen limited growth. The availability of oxygen also seems to determine the amounts of particular products excreted. For strictly anaerobic growth, Fig. 4 shows that significant quantities of ethanol are produced [41]. However, for the oxygen limited experiments described by Figs. 6 and 7, ethanol accounted for less than 3% of the xylose carbon metabolized [36].

Vollbrecht has reported that for species of *Pseudomonas*, *Paracoccus*, and *Alcaligenes*, the particular metabolites excreted depend upon the degree to which the oxygen demand of the cells is met [47]. For these species, the first product detected when respiration was systematically limited was 2-oxoglutarate. As microbial respiration was systematically forced to decrease, the order of appearance of new metabolites was 2-oxo-3-methylbutanoate, cis-aconitate, 3-hydroxybutanoate, succinate, hydrogen gas, formate, butanoate, acetoin, 2,3-butanediol, and ethanol [47]. Ethanol is only produced by these organisms at very low respiration rates. It appears that the degree of oxygen limitation plays an important role in determining which metabolites are excreted as reduced products.

For the oxygen-limited continuous cultivation of *Torulopsis utilis*, the growth rate (dX/dt) was directly proportional to the oxygen transfer rate [48]. For the same organism, the dependence of the growth rate on the dissolved oxygen concentration (DO) was a non-linear function. The oxygen supply rate may be a superior measurement to use as a growth parameter because of the direct relationship which exists between the growth rate and the oxygen supply rate, for this organism.

For *K. pneumoniae* growing in continuous culture at DO levels above 10 mm Hg, cell mass and carbon dioxide accounted for nearly 100% of the substrate carbon metabolized [49, 50]. When oxygen is available in excess, the specific oxygen uptake rate is independent of DO [49].

When the DO level is near zero, metabolism of *K. pneumoniae* depends on the actual amount of oxygen transferred from the gas phase to the cells. For a high oxygen transfer rate, acetate is the principal product [49, 50]. When the specific uptake rate decreases, butanediol becomes the predominant product [37]. Butanediol is produced over a broad range of conditions. The acetoin: butanediol ratio decreases as the oxygen uptake rate is decreased [37]. For these partially aerated conversions little ethanol is produced. However, when the oxygen supply is cut off, ethanol production increases at the expense of butanediol production. The anaerobic process by *K. pneumoniae* results in a nearly equimolar mixture of ethanol and 2,3-butanediol [41].

In summary, the oxygen supply rate appears to be important for three reasons. First, the proportions in which the various reduced products are formed is determined by the degree of oxygen limitation. Second, oxygen supply influences the respiratory quotient (which is a measure of the ratio of "fermentative" to respiratory metabolism). Finally the oxygen supply is important in determining reaction rates because it influences the cell concentration.

3.6 Water Activity

Another important variable affecting 2,3-butanediol production is water activity (which is related to osmotic pressure). The addition of salt to food acts as a preservative by retarding or preventing microbial growth. This growth inhibition is caused by a lowering of the water activity due to an increased salt concentration [51]. Water activity is an expression of the water concentration that depends on the molar concentration and the activity coefficient of each solute [52]. Increasing the solute concentration decreases the water activity of a solution.

Species of *Klebsiella* are not as osmotolerant as some other organisms [51]. Figure 8 portrays the dependence of the growth rate of *K. pneumoniae* on water activity [44]. At a water activity of 0.985, growth is only 50% optimal while at water activities below 0.975, growth rates become less than 10% optimal [44].

A decreased growth rate at lower water activities may explain why very high initial sugar concentrations are not suitable for the butanediol process.

Water activity may also explain why the butanediol process is more difficult with a complex carbohydrate source such as molasses, hydrolyzed starch, or hydrolyzed

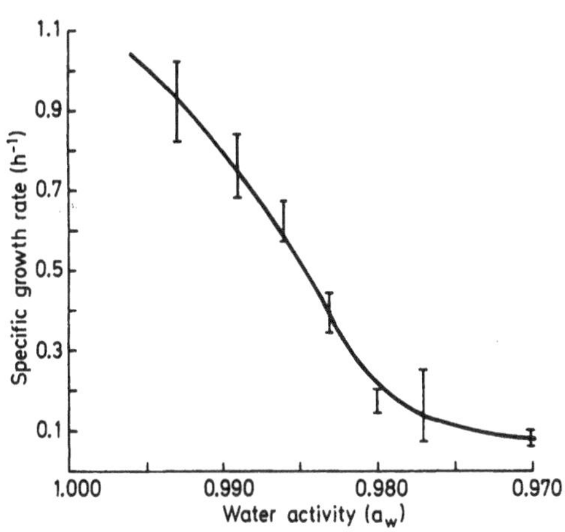

Fig. 8. Specific growth rate versus water activity [44]

cellulose. In addition to the possible presence of toxic compounds, the total solute concentration in these complex carbon sources is likely to be higher due to the presence of non-carbohydrates. This higher solute concentration causes a lower water activity which may be responsible for lower metabolic rates.

3.7 Reactor Type

Table 1 presents a comparison of some of the best results reported for various reactor systems. The performance indexes used in this comparison are final butanediol concentration, butanediol yield from sugar, and reactor productivity. Each of these may be more or less important depending on the relative costs of product recovery, carbon source, and the reactor itself.

Table 1. Performance of the 2,3-butanediol process

Process type	Substrate	Butanediol concentration[a]	Butanediol yield[b]	Butanediol productivity[c]	Ref.
Batch	Sucrose	65	0.43	1.6	53)
Batch	Xylose	30	0.31	1.3	36)
Fed-batch	Glucose	99	0.37	0.9	40)
Fed-batch	Xylose	83	0.42	1.2	36)
Immobilized cells/batch	Xylose	31	0.33	0.6	54)
Continuous	Sucrose	30	0.32	3.0	37)
Two-stage continuous	Sucrose	67	0.46	2.7	45)

[a] $g\,l^{-1}$; [b] $g\,g^{-1}$ utilized sugar; [c] $g\,l^{-1}\,h^{-1}$

Freeman obtained the maximum reported butanediol concentration (65 $g\,l^{-1}$) from batch culture [53)]. Fulmer reported a lower butanediol titer but measured a butanediol yield of 0.47 gm per g sugar [42)]. One disadvantage to batch culture is that the reactor productivity is low due to the long period of time required for the cells to accumulate to a high enough density to result in a rapid reaction rate. This productivity becomes even lower when the down-time between production runs is considered. Another disadvantage to batch culture is that the final butanediol concentration is limited by the maximum initial substrate concentration that can be tolerated by the bacteria. However, this problem can be avoided by supplying additional sugar at a slow rate during cultivation.

Highest product concentrations are obtained using a fed-batch mode. Olson and Johnson used a fed-batch system to convert a total of 266 $g\,l^{-1}$ glucose into 97 $g\,l^{-1}$ butanediol and 14 $g\,l^{-1}$ acetoin [40)]. This total product concentration of 111 $g\,l^{-1}$ is the highest reported for this conversion. However, the reactor productivity of this system is very low due to the long time required to reach the high product level (108 h).

Chambers et al., used an immobilized cell reactor to produce 2,3-butanediol from xylose in a closed-loop batch reactor [54)]. The bacterium they used (designated

AU-1-d3) was isolated from decaying wood. By immobilizing the cells on 1/2 inch Raschig rings, they were able to convert a 100 g l^{-1} xylose solution at a rate three times as great as that for their conventional batch reactor. However, the rates obtained with this organism are considerably slower than rates measured for the batch utilization of xylose by *K. pneumoniae*. In addition, oxygen transfer to the immobilized cells may be a problem.

With continuous feeding, much higher reactor productivities are possible because the reactor can be operated at steady state near the maximum possible rate. Using a dilution rate of 0.10 h^{-1}, Pirt reported a butanediol productivity of 2.7 g l^{-1} h^{-1}. Increasing the dilution rate to 0.20 h^{-1} increased the butanediol productivity to 4.6 g l^{-1} h^{-1}, but the butanediol concentration fell to 23 g l^{-1} [37]. In continuous culture the butanediol yield is usually lower because some sugar is lost in the product stream. In addition, a high final product concentration may be more difficult to attain by continuous culture because the entire reactor is continually subject to product inhibition to the maximum extent.

Pirt designed a remarkably efficient system for producing 2,3-butanediol by using a two-stage continuous reactor. This system produced a high product concentration in high yield and at a rapid rate [45]. However, his reported yield may be exaggerated because a substrate carbon recovery of over 100% indicated some analytical errors in this study. Furthermore, the yield is based on utilized sugar only and does not account for sugar lost in the effluent stream. Nevertheless this two-stage continuous system demonstrated superior overall performance compared to the other reactor types.

4 Conclusion

Lignocellulosic materials represent a large untapped reservoir of carbohydrates which are available for conversion into more valuable products. *Klebsiella pneumoniae* can convert all the sugars contained in hemicellulose and cellulose hydrolysates to 2,3-butanediol. Of particular note is the ability of this organism to convert xylose to butanediol in the same yield (and at nearly the same rate) as it converts glucose to butanediol. The kinetics of butanediol production from xylose and glucose have been examined. Butanediol concentrations as high as 99 g l^{-1} have been obtained with this organism.

Future work will likely entail optimizing the production of 2,3-butanediol from actual hemicellulose or cellulose hydrolysates. Osmotolerant strains, which grow better in hydrolysates having a lower water activity, may increase conversion rates. In addition, further improvements in reactor design, such as the two-stage continuous system or the immobilized cell system, may enhance the efficiency of butanediol production.

5 Acknowledgment

The authors wish to acknowledge the support from the National Science Foundation, Grant No. 8022426A1.

6 References

1. Tsao, G. T.: Process Biochemistry *13* (10), 12 (1978)
2. Burwell, C. C.: Science *199*, 1041 (1978)
3. Tsao, G. T., et al.: Annual Reports on Fermentation Processes. (D. Perlman, G. T. Tsao eds.), *2* (1), New York: Academic Press 1978
4. Tyner, W. E.: Biotech. Bioeng. Symp. *10*, 81 (1981)
5. Lipinsky, E. S.: Proc. 3rd Annual Biomass Energy Systems Conf., 229, Solar Energy Research Inst., Dep. of Energy, Golden, CO 1979
6. Stephens, G. R., Heichel, G. H.: Biotech. Bioeng. Symp. *5*, 27 (1975)
7. Sloneker, J. H.: ibid. *6*, 235 (1976)
8. Ladisch, M. R.: Process Biochem. *14* (1), 21 (1979)
9. Dunning, J. W., Lathrop, E. C.: Ind. Eng. Chem. *37*, 24 (1945)
10. Leonard, R. H., Hajny: ibid. *37*, 390 (1945)
11. Flickinger, M. C.: Biotech. Bioeng. *22* (1), 27 (1980)
12. Flickinger, M. C., Tsao, G. T.: Annual Reports on Fermentation Processes. (D. Perlman, G. T. Tsao eds.), *2* (2), New York: Academic Press 1978
13. Palson, B. O., et al.: Science *213*, 513 (1981)
14. Villet, R. H.: Chem. Eng. Progress *77*, 59 (1981)
15. Gong, C. S., et al.: Adv. Biochem. Eng., (A. Fiechter ed.), *20*, 93 Berlin: Springer 1981
16. Rosenburg, S. L.: Enzyme Microb. Technol. *2*, 185 (1980)
17. Emerson, R. R.: M. S. Thesis, Purdue Univ. 1981
18. Morell, S. A., Geller, H. H., Lathrop, E. C.: Ind. Eng. Chem. *37*, 877 (1945)
19. Ledingham, G. A., Neish, A. C.: Industrial Fermentations. (L. A. Underkofler, R. J. Hickey eds.), *2*, New York: Chemical Publishing Co. 1954
20. Anderson, R. L., W. A. Wood: J. Biol. Chem. *237*, 296 (1962)
21. Buchanan, R. E., Gibbons, N. E.: Bergey's Manual of Determinative Bacteriology, Eighth Ed., p. 294, Baltimore: Williams and Wilkins 1974
22. Yu, E. K. C., Saddler, J. N.: Biotech. Lett. *4*, 121 (1982)
23. Mortlock, R. P., Wood W. A.: J. Bact. *88*, 838 (1964)
24. Mortlock, R. P., Wood, W. A.: ibid. *88*, 845 (1964)
25. Mortlock, R. P., Fossitt, D. D., Wood, W. A.: Proc. Nat. Acad. Sci. *54*, 572 (1965)
26. Barnett, J. A.: Adv. Carbohydr. Chem. Biochem. *32*, 125 (1976)
27. Wood, W. A.: in: The Bacteria Volume II Metabolism. (I. C. Gunsalus, R. Y. Stanier eds.), New York: Academic Press 1961
28. Neish, A. C., Simpson, F. J.: Can. J. Biochem. Physiol. *32*, 147 (1954)
29. Altermatt, H. A., Simpson, F. J., Neish, A. C.: ibid. *33*, 615 (1955)
30. Juni, E.: J. Biol. Chem. *195*, 715 (1952)
31. Johansen, L., Bryn, K., Stormer, F. C.: J. Bact. *123*, 1124 (1975)
32. Stormer, F. C.: J. Biol. Chem. *243*, 3735 (1968)
33. Loken, J. P., Stormer, F. C.: Eur. J. Biochem. *14*, 133 (1970)
34. Larsen, S. H., Stormer, F. C.: ibid. *34*, 100 (1973)
35. Voloch, M.: Ph. D. Thesis, Purdue Univ. 1981
36. Jansen, N. B.: Ph. D. Thesis, Purdue Univ. 1982
37. Pirt, S. J., Callow, D. S.: J. Appl. Bact. *21*, 188 (1958)
38. Topiwala, H., Sinclair, C. G.: Biotech. Bioeng. *13*, 795 (1971)
39. Esener, A. A., Roels, J. A., Kossen, N. W. F.: ibid. *23*, 1401 (1981)
40. Olson, B. H., Johnson, M. J.: J. Bact. *55*, 209 (1948)
41. Neish, A. C., Ledingham, G. A.: Can. Res. *278*, 694 (1949)
42. Fulmer, E. I., Christensen, L. M., Kendall, A. R.: Ind. Eng. Chem. *25*, 798 (1933)
43. Long, S. K., Patrick, R.: Adv. Appl. Microb. *5*, 135 (1963)
44. Esener, A. A., et al.: Advances in Biotechnology (Proc. VI Int. Ferm. Symp.) *1*, 339, Ontario: Pergamon Press 1981
45. Pirt, S. J., Callow, D. S.: Selected Sci. Papers from the 1st Sup. di Sanita, *2*, 292 (1959)
46. Paca, J.: Folia Microbiol. *23*, 108 (1978)
47. Vollbrecht, D., ElNawaway, M. A.: Eur. J. Appl. Microbiol. *9*, 1 (1980)

48. Button, D. K., Garver, J. C.: J. Gen. Microbiol. *45*, 195 (1966)
49. Harrison, D. E. F.: ibid. *46*, 193 (1967)
50. Pirt, S. J.: ibid. *16*, 59 (1957)
51. Scott, W. J.: Aust. J. Biol. Sci. *6*, 549 (1953)
52. Pirt, S. J.: Principles of Microbe and Cell Cultivation, p. 147, New York: J. Wiley and Sons 1974
53. Freeman, C. G., Morrison, R. I.: J. Soc. Chem. Ind. Lond. *66*, 216 (1947)
54. Chambers, R. P., Lee, Y. Y., McCaskey, T. A.: Proc. 3rd Annual Biomass Energy Systems Conf., p. 255, Solar Energy Research Inst., Dep. of Energy, Golden, CO 1979

Bacterial Conversion of Pentose Sugars to Acetone and Butanol

B. Volesky and T. Szczesny
Department of Chemical Engineering, McGill University, Montreal, Canada H3A 2A7

Metabolic activities of some anaerobic *Clostridia* strains result in the production of solvents, acetone, butanol, ethanol, iso-propanol etc., and hydrogen gas, all of interest to industry. Hexose sugars are readily utilized for this purpose, so are hexose- and pentose-based polysaccharides. Pentoses can also be utilized, however, the production rates observed are somewhat lower. Revived interest in the process is currently gaining momentum and new techniques of genetic manipulation of microorganisms coupled together with novel bioprocess technology, solvent recovery techniques, and minicomputer process control/optimization, offer a promise of resulting in an efficient and competitive process modification. Wood, agricultural residues and waste liquors constitute a cheap source of renewable raw materials for industrial solvent production.

1 Introduction

The current rate of consumption of hydrocarbon feedstocks and recent experience with acute shortages of this most important and strategic raw material contribute to the realization that it is a resource which sooner or later will run out. In an intensified search for alternatives, considerable attention currently is being paid to the development of processes based on renewable natural resources and to the investigation of unconventional energy sources. This trend has brought about development of a number of new biotechnology based and energy saving processes, and some of the conventional approaches are being reexamined in light of contemporary advanced biotechnology and efficient process optimization techniques.

1.1 Pentoses from Natural Sources

Wood and agricultural by-products are considered to be the most abundant renewable resource materials. Based on photosynthetic processes, they are virtually inexhaustible. Wood is composed of three components — cellulose, hemicellulose and lignin. The cellulose and the hemicelluloses are deposited in the secondary wall of the cells. Cellulose in the form of long part-crystalline micro-fibrils is surrounded by the amorphous hemicellulose; and the whole is embedded in a matrix of lignin [1,2,3].

Cellulose, the chief constituent of the cell wall of all land plants is a linear polymer composed of D-glucose units linked by β-glycosidic bonds, and in its native state has a weight-average degree of polymerization in the region of 10.000. In most cases unlike cellulose, the hemicelluloses are heteropolymers containing D-xylose, D-glucose, D-mannose, D-galactose, L-arabinose, 4-0-methyl-D-glucuronic acid and occasionally a few other sugars. The more common hemicelluloses usually contain xylose or mannose and glucose as the main components and closely resemble cellulose in the manner of linkage; although, they have a much lower degree of polymerization (about 50–400). In hardwoods the main hemicellulose present is usually a β (1–4)- linked xylan, containing side chains of 4-0-methyl-D-glucuronic acid, whereas in softwoods the predominant hemicellulose is usually a glucomannan, a linear polymer of (1–4)-β-linked glucose and mannose. The number of hemicelluloses now characterized indicate a possible relationship between molecular structure and botanical origin [1,2]. The structures of the hemicelluloses from many deciduous and coniferous trees as well as from agricultural by-products have been reported.

Pectic polysaccharides, mainly composed of D-galacturonic acid and L-arabinose, also occur widely, but in woods these are evenly distributed and are not usually isolated in the hemicellulose fraction.

From many published studies, certain generalizations have emerged, one of the most notable being that hemicelluloses from coniferous trees have a high mannose content, whereas those from the deciduous trees are low in this sugar and possess a high xylose content [2,3,4].

Pulp and paper making processes, using wood as raw material, result in a variety of wastewaters containing carbohydrates originating from wood chip hydrolysis. Development of a practical method of disposal and/or purification is a problem in itself, while the utilization of the potential values contained in this material has for long been of interest. Regardless of the upstream processing, the wood carbohydrates, being of renewable origin, hold a remarkable promise as raw materials for numerous types of microbial processes and the whole area has been lately receiving increasing attention. In general, hydrolysis of wood components, whether enzymatic, acid, alkaline, or combined with physical means, results in some lignin and a variety of oligosaccharides and simple pentose and hexose sugars. While hexoses and glucose in particular serve as most suitable substrates for microbial activity, pentoses are somewhat less favoured by microorganisms.

Some strains of yeasts, however, have been successfully used in pentose conversion to chemical products such as ethanol [5,6] or high-protein biomass: Single-Cell Protein [7,8]. Anaerobic microbial processes have received particular attention for their potential to produce fuel compounds (alcohols, hydrogen), alternative chemical feedstocks and solvents. While only a few strains of yeasts have been reported to be capable of anaerobic biosynthesis of the above-mentioned types of compounds, many bacteria are capable of using pentoses, and therefore are of considerable interest.

2 Acetone-Butanol Producing Microorganisms

The discovery of normal butyl alcohol as a regularly occurring constituent of fusel oil in 1852 was attributed to Wirtz. Ten years later, 1862, Pasteur [9], from his experiments, concluded that butyl alcohol was a direct product of anaerobic conversion of lactic acid and calcium lactate. Between 1876 and 1910, several scholars researched the production of acetone-butanol and related solvents. Some of the more prominent individuals include Beijerinck (1893) [10]; Duclaux (1895) [11]; Emmerling (1897) [12]; Fitz (1882, 1884) [13,14]; Grassberger and Schattenfroh (1902) [15]; Grimbert (1893) [16]; Gruber (1887) [17]; Schardinger (1905) [18] and Winogradsky

Table 1. *Clostridia* capable of utilizing pentose sugars

Microorganism	Pentose		
	Arabinose	Ribose	Xylose
Group I:			
C. butyricum	(d)	+	+
C. beijerinckii	+	—	+
C. oroticum	+	+	+
C. rubrum	+	(d)	+
C. fallax	—	(d)	+
C. pasteurianum	+	(d)	—
C. tyrobutyricum	—	—	+
Group II:			
C. bifermentans	—	(d)	(d)
C. sordellii[a]	(d)	(d)	(d)
C. acetobutylicum	+	—	+
C. felsineum	+	—	+
C. chauvoei[a]	—	+	—
C. difficile[a]	(d)	(d)	+
Group III:			
C. sphenoides	(d)	(d)	+
C. indolis	(d)	(d)	+
C. scatologenes	+	+	+
C. tertium	(d)	+	+
C. sartagoformum	—	(d)	+
C. cellobioparum	+	+	+
C. thermosaccharolyticum	+	(d)	+
C. pseudotelanicum	+	+	+
C. glycolycum	—	—	+
C. barkeri	—	+	+
C. perenne	—	+	—

[a] — pathogenic for laboratory animals
(d) — reactions differ

(1902)[19]. In 1912, Weizmann[20] reported to have discovered a bacterial culture with outstanding capabilities for anaerobic conversion of grains such as corn into acetone and butanol. World War I contributed to a rapid development of the microbial process on an industrial scale. In acetone-butanol production, and to a lesser extent ethanol, the bacterial strains mostly involved are *Clostridium* species. The genus is divided into 5 groups containing altogether 61 reference strains. The division into groups is with respect to organisms' ability to hydrolyze gelatin (Group II) or not (I) and the location of the endospore within the vegetative cell (subterminal: Groups I and II; terminal: Groups III and IV). Group V is distinguished by organisms' nutritional requirements. Solvent producing *Clostridia* can be found in any of the five groups. They are rods, usually motile by means of a peritrichous flagella; occasionally non-motile. They form ovoid to spherical spores that usually distend the bacilli. At least in the early stages of growth they are generally gram positive. Being chemo-organotrophs, they metabolize sugars, poly-acohols, amino acids, organic acids, purines and other organic compounds under strict anaerobic conditions. They do not reduce sulphate and some species can fix nitrogen. While a number of *Clostridium* strains are capable of utilizing pentose sugars as carbon and energy sources the rate and efficiency of the pentose metabolism differs markedly when compared to that of hexose. Table 1 lists those *Clostridia* indicated in Bergey's Manual[21] as capable of utilizing three common pentose sugars.

It is noteworthy that Bergey's Manual does not list a number of *Clostridium* species of industrial importance which have either been mentioned in the relevant literature[22] or which are key to a patented solvent-production process. Anaerobic biosynthesis of acetone-butanol was the first large-scale microbial process in which the exclusion of other kinds of microorganisms from the culture vessel became a factor of major importance for the success of the operation. The medium used for the process employing most often *Clostridium acetobutylicum* or *Clostridium butylicum* is also favourable for the development of lactic acid bacteria whose activity is detrimental to the process. A more serious problem stems from *Clostridia* being rather susceptible to a phage attack. The clostridial phage strain specificity, however, enables to minimize the process culture losses by production strain rotation, coupled with efficient sterilization techniques, clean operation and careful stock culture maintenance.

3 Biochemistry of Pentose Utilization

The anaerobic biochemical reaction sequence degrading hexose sugars is based on the well known glycolysis pathway (Embden-Meyerhof pathway) 1 resulting in formation of pyruvate and energy in the form of ATP. The reactions proceeding from glucose to pyruvic acid occur in a wide variety of microorganisms, but the resulting pyruvate may be processed further in a number of different ways. Among the end-products of some strains of *Clostridia* are acetone, butanol, ethanol and iso-propanol, fermented with acetate and butyrate as major intermediates, while gaseous CO_2 and H_2 are given off. The anaerobic catabolism of pentoses, according to the theories generally adopted[23], can proceed along two major routes:

a) by fission between the carbons 2 and 3 of pentoses with formation of glycoaldehyde and glyceraldehyde or their phosphorylated derivatives [24,25,26] as seen in Fig. 1a.
b) by pentose phosphate pathway leading to a formation of a hexose phosphate or other intermediate compounds capable of entering the glycolytic pathway [27,28].

The latter scheme makes use of transaldolase and transketolase enzymes catalyzing the interconversion of three-, four-, five- and six-carbon sugars (Fig. 1b). It allows pentoses to be used as energy sources by microorganisms that lack the enzyme phosphoketolase which is required for the former scheme. Indirect evidence of this scheme was obtained from investigation of *Lactobacillus pentoaceticus* growing on C^{14}-labeled D-arabinose and D-xylose [29,30]. All radioactive atoms were found in the methyl carbon of acetic acid and not in the carbons of lactic acid, which accumulated in equimolar amounts. This result proves that the methyl- and carboxyl-groups derive respectively from carbons 1 and 2 of arabinose and xylose suggesting that a ketopentose or a similar phosporylated derivative is formed as an intermediate. Experiments carried out with *Clostridium acetobutylicum* positively identified triose-phosphates and pyruvic acid but fail to confirm the presence of glycoaldehyde and acetaldehyde [31]. The former compound, however, was isolated during the anaerobic bioconversion of pentoses by *Acetobacter acetizenum* [32].

In general, the metabolic activity is aimed at the production of pyruvate as a key intermediate compound. Energy production for cellular anabolic processes is the main purpose of the whole metabolic activity. Strict anaerobes control their

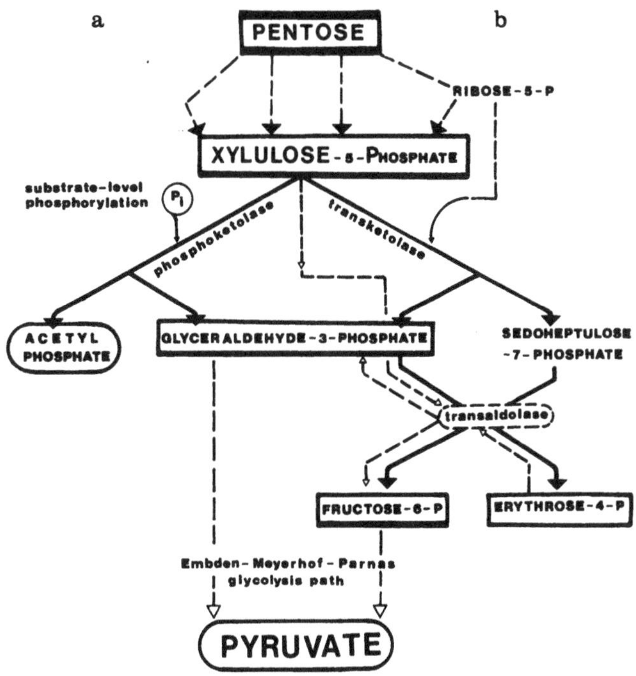

Fig. 1. Pentose metabolism by anaerobic bacteria. **a** phosphoketolase — catalyzed fission of xylulose-5-P between the carbons 2 and 3; **b** pentose phosphate pathway using transketolase leading to a formation of a hexose phosphate

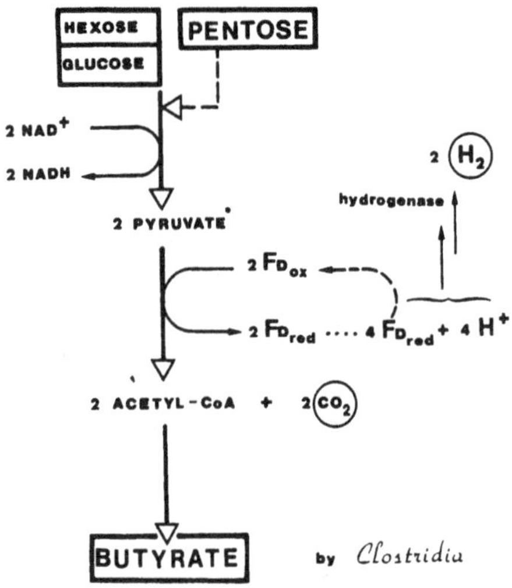

Fig. 2. Hydrogen production by *Clostridia* is mediated by ferredoxin upon oxidizing pyruvate

high-energy ATP production by controlling the flow of electrons in their metabolism of substrate, such a mechanism is to dispose of excess electrons in the form of molecular H_2 through the activity of the hydrogenase enzymes [33]. From practical point of view this step is important since it is responsible for generation of hydrogen gas (Fig. 2), a very useful product and an added benefit in the solvent biosynthesis process. The electrons are transferred to electron carriers such as NAD^+ or FAD, and the reoxidation of these carriers requires another molecule as electron acceptor. Saccharolytic *Clostridia* couple the oxidation of pyruvate to acetyl-CoA and CO_2 with the reduction of ferredoxin. The redox potential of this coenzyme is nearly —400 mV and it is low enough to allow the reduction of protons to molecular hydrogen. It appears that ferredoxin and CoA are functioning in unison in enabling pyruvate dehydrogenase to act. There is evidence [34] that ferredoxin is the limiting factor in pyruvate oxidation. In irondeficient environment, there is no hydrogen formation. In all saccharolytic *Clostridia* absence of H_2 evolution leads to a shift to lactate production. Such is the case in the absence of dehydrogenases. In the metabolic pathway, the production of acetate as a sole end-product would not be satisfactory since it becomes progressively more difficult to reoxidize the reduced $NADH + H^+$ to NAD^+ with the decrease of the pH into the acid region. Therefore, it is not surprising to find a cyclic mechanism that forms butyrate which is much less acid end-product than acetate. Consequently, as soon as butyrate lowers the pH to $\sim$4.0 the system favours the production of acetone and concurrently converts the butyrate to butanol (Fig. 3).

Clostridium acetobutylicum possesses a transferase system which diverts acetoacetyl — CoA from the normal cyclic mechanism to produce acetoacetate, which undergoes decarboxylation to acetone. This final step is irreversible. *C. butylicum* culture is capable of synthesizing iso-propanol by this scheme. Diversion of the original cycle system to form acetone stops further production of butyrate. Since this eliminates also two NAD^+ generating steps, some other reduction processes must be found:

thus butanol is produced. The reduction of butyrate to butanol takes place in three consecutive steps. There is an alternative to this step requiring ATP and CoA. This production of butanol occurs only after the change to the production of acetone has taken place. Ethanol is a by-product of the major pathways. The branching point is Acetyl-CoA. The general metabolic scheme is depicted in Fig. 3.

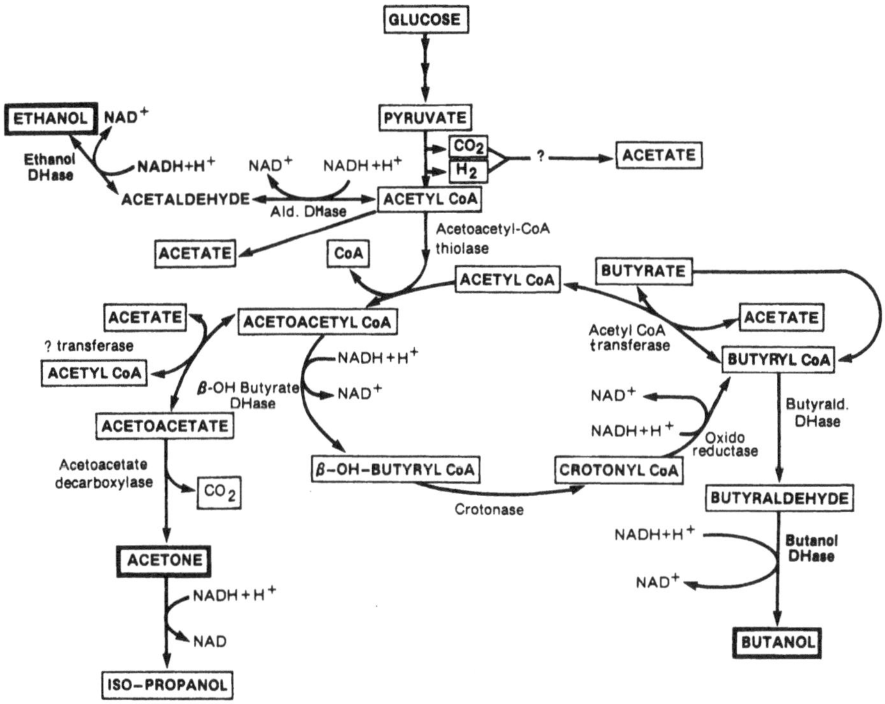

Fig. 3. General anaerobic metabolic scheme leading to the microbial production of solvents, CO_2 and hydrogen gas

Control of acetate production is a key point in the control of the whole enzymatic sequence. This is in fact so since all the end-products are comprised of acetate units (butanol, acetone, ethanol). Hence, the availability of acetate is a very important factor in the anaerobic microbial process. Although only two moles of acetate are derived from one mole of hexose sugar (glucose) it has been found [35] that in actual fact there are three moles of acetate formed by *C. thermoaceticum*, the third one derived from CO_2. One or more of the enzymes responsible for this conversion are inactivated at lower temperatures and exposure to 57 C (*C. acetobutylicum* to 45 C) favoured reactivation. Vitamin B_{12} was found to be essential [35].

Despite early large-scale utilization of the acetone-butanol microbial production process, the exact nutrient requirements of production *Clostridium* strains have never really been ascertained. More often than not additions of various organic materials to the basic medium have been widely practiced supplementing it with organic nitrogen, amino acids, vitamins and other growth factors of undefinable nature. The main substance for the large-scale process, when it operated decades ago, was usually a very complex mixture of carbohydrates and other carbonaceous and inorganic compounds coming from natural sources more or less pre-processed which differed greatly from one operation to another. Understandably, this was so for the reason of production economy. The variety of substances is reflected in the laboratory studies reported in the literature. In the following paragraphs the focus will be on the work carried out with acetone-butanol producing strains of *Clostridium* growing on media containing pentose sugars.

4 Pentoses in Acetone-Butanol Production

An attractive feature of the solvent producing *Clostridia* is their capacity to utilize pentose sugars and certain complex polysaccharides. This makes them doubly interesting for exploitation in production of cheap solvents from wood, wood waste and agricultural residues. Early investigators of such processes focused on these substrates in the period between the two world wars and further work was being published in the fifties.

Table 2. Effect of different methods of pretreating waste sulfite liquor prior to fermentation. After Wiley et al. 1941 [40]

Expt. No.	Pretreatment	Medium No.	Organism[d] No.	% Sugars utilized	Fermentation time, days
1	Boiling, neutralization	1[a]	39	None	10
			04	None	10
2	Boiling, neutralization, dilution to 40% of original strength	1[a]	39	62.7	7
			64	63.7	7
3	Boiling, neutralization, clarification with norite	1[a]	39	None	10
			64	None	10
4	Boiling of SO_2 at pH 1.5 neutralization	1[a]	39	70.0	7
			64	None	7
5	SO_2 pptn. at pH 10.0 by $Ca(OH)_2$ neutralization	2[b]	39	74.3	3
6	SO_2 pptn. at pH 10.0, lignin pptn. at pH 11.5 (room temp) neutralization	2[b]	39	72.6	2
7	SO_2 pptn. at pH 10.0, lignin pptn. at pH 11.5 Ca pptn. with 1% Na_2SO_4 before neutralization	3[c]	39	74.9	2

[a] 3% malt sprouts, 0.3% NaH_2PO_4, 0,2% $CaCO_2$ in experiments 1, 2, 3 and 0.1% $CaCO_2$ in experiment 4;
[b] 0.5% $(NH_4)_2HPO_4$, 0.1% molasses, and 0.1% $CaCO_2$;
[c] 0.05% $(NH_4)_2HPO_4$, 0.1% molasses, and 0.1% $CaCO_2$;
[d] organisms: No. 39 — *Cl. butylicum* (Fitz)
No. 64 — *Cl. butylicum* (Wisconsin)

Pulp and paper industry waste waters, namely waste sulfite liquors, present one of the greatest industrial waste problems while the utilization of the potential values contained in this materials has attracted considerable attention. The carbohydrate fraction of these liquors containing hexose and pentose sugars can be utilized by microbes and apart from single-cell protein production process based on aerobic Torula yeast, anaerobic ethanol production by yeast as well as bacterial acetone-butanol synthesis was utilized, reflected in patent coverage of process alternatives [36, 37, 38 39]. Having as an added benefit a substantial waste BOD removal, the process based on waste sulfite liquor was not without problems, however [40]. The liquor, usually at low pH of 2.5–3.5, with ~10% solids and total reducing sugars expressed as glucose at up to ~3%, contains high concentrations of free and loosely bound sulfur dioxide, calcium in an exchangeable form, formic and oxalic acids, lignin, and furfural which exhibit more or less toxic effect on microorganisms. Many experiments were conducted to determine the best methods of pretreating the sulphite liquor to render it more easily available to microbes (Table 2). Following the pretreatment which can remove up to 45% of the BOD and supplementing the nitrogen and phosphorus content of the liquor, its degradability by *C. butylicum* (Fitz) may be in excess of 80% (based on total sugars). With the final BOD lowered to 20% of its original value, solvent yields ranged around the average of 30%. The total solvents contained typically 75% butanol, 20% acetone and 5% ethanol [40,41,42]. In order to achieve higher accumulation of solvents, higher sugar levels are desirable, however, attempts to use liquors with more than 15% solids and 3% total sugars generally failed. Validity of the individual experimental results is very difficult to extend because of the widely variable nature of waste liquors originating from different mill operations processing different types of wood. The solvent process based on waste sulfite liquor was considered feasible and was being utilized on a large scale when market conditions warranted it. Novel pulp and paper making technology has been making its impact and the traditional types of waste liquors are becoming less common as the older mills are being phased out. The general trend is obviously to close the water circulation within the pulping process, minimizing thus the waste-water problem with effluents also changing their character. Wood and agricultural residues as a plentyful source of biodegradable carbohydrate materials, however, were receiving considerable attention in the post-war decade. This trend, somewhat suppressed during the decades of abundant and cheap hydrocarbon feedstocks, has recently been strongly revived. The early efforts [43], based on good microbial solvent producing strains of *C. butylicum* originated in Wisconsin, indicated the problems associated with bioconversion of wood hydrolyzates into solvents. In some cases the inhibitory effect was accounted for by the presence of xylose decomposition product furfural [44] prompting its removal together with some other unidentified inhibitors, by steam stripping which appeared the simplest. Hydrolyzates, lime-neutralized to pH 6.5, containing more than 3% sugars gave a lesser extent of the bioconversion with increasing period of wood hydrolysis. This solvent yields ranged from 25% to 38% of the sugar utilized. Hydrolysates from different types of wood contained different levels of pentoses and inhibitors. Pure xylose is not so easily utilized by microbes as glucose or glucose-xylose mixture and usually gives acids rather than neutral solvents. While the pretreatment of hydrolyzates with active carbon and the presence of reducing agents in the bioprocess was beneficial [43,44,45] it might not be essential[39,46]. Charcoal treatment was ineffective in improving the fermentation of pure xylose and detrimental effect of copper ions was clearly demonstrated[45]. Somewhat contradictory opinions exist concerning the optimal levels of pentoses in hydrolysate fractions removed early in the hydrolysis process. 70% pentose content was reported as the optimum[39]. This patented microbial process did not benefit from addition of hexoses or hexose-containing mash as indicated by others[45,46,47] and a 50/50 mixture of wood xylose and hexoses at pH 6.5 gave 1.7% butanol, 0.9% acetone and 0.19% ethanol accumulation at the end of a 48 h fermentation. Mild hydrolysis conditions seem to be favourable for further microbial process and so are the precipitated lignin particles[48] present in the neutralized liquor. The solvent yields by *C. butylicum* (Prazmowski) used in the Latvian work ranged from the usual 30% to 38% of available sugars while butanol fraction constituted 67% of total solvents.

After all this work, however, a question still remains concerning the quantitative aspects of degradability of not only individual pentose sugars but also some hexoses. Also the effect of sugar levels and of mixing hexoses with pentoses on the solvent culture performance remains somewhat unclear.

Recent wirk of Compere and Griffith[49] partially addresses this void. The authors investigated some high-solvent producing clostridia, namely two *C. acetobutylicum* strains (NRRL B527 and B3179), two strains of *C. butylicum* (NRRL B592 and

Table 3. *Clostridium butylicum* on pentoses and hexoses

Clostridium butylicum NRRL B 592

% w/v Carbohydrate	% v/v Total solvents			Solvent distribution B:A:E % tot. solvents		
	2.5	5	10	2.5	5	10
Xylose	0.09	0.54	0.08	44:12:44	66:18:16	38:38:24
Arabinose	trace	1.62	1.74	trace:trace:trace	81:12: 7	79:17: 4
Xylan	0.16	0.32	2.59	62:12:26	81: 6:13	85:14: 1
Glucose	1.20	1.62	1.98	73:14:13	72:17:11	62:26:12
Cellobiose	0.72	2.09	2.87	72:21: 7	73:21: 6	53:22:25
Dextrin	0.84	1.25	4.08	78:14: 8	81:12: 7	75:21: 4

Clostridium butylicum NRRL B 593

% w/v Carbohydrate	% v/v Total solvents			Solvent distribution B:A:E % tot. solvents		
	2.5	5	10	2.5	5	10
Xylose	0.28	0.25	0.01	75:13:12	76:12:12	trace:trace:100
Arabinose	0.21	0.34	0.17	67: 9:24	73: 6:21	88: 6: 6
Xylan	0.17	0.40	1.10	59:18:23	82: 5:13	85:12: 3
Glucose	1.23	1.07	2.14	85:12: 3	76:14:10	69:22: 9
Cellobiose	0.57	1.12	1.91	89: 5: 6	78:15: 7	74:21: 5
Dextrin	0.82	1.04	1.85	82:10: 8	79:12: 9	93: 2: 5

Modified from Compere and Griffith, 1979[49]

B593), a strain of *C. pasteurianum* (NRRL B598) and a *Clostridium* mixture with added *Klebsiella pneumoniae* (NRRL B427). The latter mixed culture as well as another one with a kefir yeast in addition to the previous multi-strain system did not exhibit any significant activity in solvent production compared to pure cultures particularly relative to butanol accumulation. The results for pure *Clostridia* cultures, however, are difficult to quantitatively compare and there is not statistical significance attached to the experimental values reported by Compere and Griffith [49], whereby a serious shortcoming is the lack of data indicating the degree of sugar utilization. Some conclusions can be drawn concerning comparison of pentose and hexose substrates (Tables 3 and 4):

a) pentoses result in much smaller amounts of solvents produced with growth and solvent synthesis rates substantially lower;
b) disaccharides and oligosaccharides give higher solvent accumulation than monosaccharides;
c) the production culture(s) may be more sensitive to higher pentose concentrations than the case is for hexoses whereby even 10% substrate concentrations resulted in good solvent production. The culture of *Clostridium butylicum* NRRL B592 may be particularly efficient in solvent production from wood hydrolyzates containing xylan and cellobiose, and dairy wastes, which contain lactose. It would also be useful for the anaerobic bioconversion of starch substrates.

Table 4. *Clostridium acetobutylicum* on pentoses and hexoses

Clostridium acetobutylicum NRRL B 3179

% w/v Carbohydrate	% v/v Total solvents			Solvent distribution B:A:E % tot. solvents		
	2.5	5	10	2.5	5	10
Xylose	0.13	1.22	0.02	46: 7:47	47:47: 6	trace:50:50
Arabinose	0.11	1.34	0.49	54: 9:37	47:47: 6	77:16: 7
Xylan	0.04	0.19	0.04	25:25:50	37:36:27	50:25:25
Glucose	0.20	1.15	0.04	75:15:10	47:47: 6	50:25:25
Cellobiose	0.59	1.41	2.45	86: 7: 7	47:47: 6	79:20: 1
Dextrin	0.42	1.75	0.03	78:12:10	45:45:10	33:34:33

Clostridium acetobutylicum NRRL 527

% w/v Carbohydrate	% v/v Total solvents			Solvent distribution B:A:E % tot. solvents		
	2.5	5	10	2.5	5	10
Xylose	0.38	0.96	0.10	79: 5:16	83: 9: 8	50:20:30
Arabinose	0.62	0.33	0.51	80:11: 9	76:12:12	75:14:11
Xylan	0.20	0.16	0.03	65:10:25	56:13:31	trace:33:67
Glucose	0.42	0.69	0.05	55:19:26	75:12:13	trace:60:40
Cellobiose	1.22	0.26	0.14	71:22: 7	50:12:38	36:21:43
Dextrin	0.80	0.68	0.06	79:14: 7	76:10:14	33:34:33

Modified from Compere and Griffith, 1979 [49]

Confirmation of the above statements can be found in the older tests with xylose [45,50] and arabinose [50] as well as in more recent verifications [51,52,53]. The relevant results are presented in comparative Tables 5 and 6 with original results in Table 7. Exact comparison of results is somehow difficult to make because experiments were carried out with different culture strains, on different media, and other culture parameters also differed to a certain degree. The trends and indications, however, can be discerned.

An interesting approach was briefly reported on recently [54] in which *C. acetobutylicum* was grown on pentoses (arabinose or xylose) in a mixed "consecutive" culture with *Saccharomyces cerevisiae*. While the yeast utilized hexose sugars from molasses (5% solids) producing ethanol anaerobically (22 $g\,l^{-1}$ in 48 h), the bacterium was added into the culture after 24 h and converted arabinose and xylose component of the medium (30 $g\,l^{-1}$) into 6.6 $g\,l^{-1}$ and 3.7 $g\,l^{-1}$ respectively of butanol, in another 170 h. The higher butanol accumulation from arabinose reflects higher conversion rates observed with this pentose sugar. The culture time, however, appears to be unpractically long and the complete utilization of pentoses, assumed by authors for the solvent yield calculation was not analytically verified. In the course of this study, the extent of butanol biosynthesis inhibition by *C. acetobutylicum* was reconfirmed.

Table 5. *Clostridia* growing on xylose-glucose mixtures[a]

Ref.	Xylose %								
	100	90	80	70	60	50	25	0	Comments
[45]	41.0	64,4	61.3	82	86.5	82.1	89.2	88.6	*Cl. butylicum* (no time given)
Authors	46.0	23.2	30.4	—	50.0	56.0	—	80.0	*Cl. acetobutylicum* (76 h)
[44]	49	—	—	—	—	95	—	98	*Cl. butylicum* (65 h)

[a] approximately 5% initial sugar concentrations

Table 6. Comparison of different microbial utilizations of pentoses

Sugar		Total solvents % v/v			Solvent distribution *B:A:E % tot. solvents			Ferm. time h	Ref.
Init. %	Used %	Xyl	Ara	Glu	Xyl	Ara	Glu		
2.07	94	**—	0.52	0.52	—	48:39:13	60:28:12	84	[25]
2.00	95	0.60	0.36	0.64	61:29:10	47:43:10	59:31:10	81	[50]
6.25	—	—	—	—	55:30:15	50:39:11	65:25:10	80	[50]
5.03	89	1.38	—	—	62:32: 6	—	—	—	[45]
5.0	46	0.73	—	1.12	67:30: 3	—	65:31: 4	72	Authors

* B — butanol; A — acetone; E — ethanol; ** (—) means "data not available"

Table 7. *Clostridium acetobutylicum* on Xylose and Glucose

Initial mix of sugars $g\,l^{-1}$		Sugars utilized				TNVP $g\,l^{-1}$	Solvent distribution % TNVP			Acetic acid $mg\,l^{-1}$	Butyric acid $mg\,l^{-1}$
		$g\,l^{-1}$	%	%	%						
Glu	Xyl	Total	Total	Glu	Xyl		BuOH	Acetone	EtOH		
5	45	11.6	23.2	100	14.6	1.6	65.1	30.0	4.9	2937	2710
10	40	15.2	30.4	100	13.0	2.6	70.8	25.4	3.8	2146	2087
20	30	25.0	50.0	100	16.7	6.6	70.4	25.4	4.2	2344	953
25	25	28	56.0	100	12.0	9.5	67.0	29.0	4.0	1714	480
—	50	23	46.0	—	46.0	7.3	66.6	30.0	3.4	2120	1347
50	—	40	80.0	80	—	11.2	65.0	31.1	3.9	2000	956

Medium: sugar 50 $g\,l^{-1}$, ammonium sulphate 9.6 $g\,l^{-1}$, ferrous sulphate 0.25 $g\,l^{-1}$,
(1 l) difco yeast extract 7.5 $g\,l^{-1}$, calcium phosphate 0.22 $g\,l^{-1}$, magnesium sulphate 1.0 $g\,l^{-1}$

Simultaneously, an attempt was reported [55] on bioconversion of agro-wastes into acetone and butanol by *C. saccharoper butylacetonicum* [56]. Following a 2.25% alkali autoclaving pretreatment of bagasse or rice straw, a 30 h enzymatic hydrolysis (*A. wentii* and *T. reesei* culture filtrate) resulted in 6% reducing sugar content of a 10% slurry. Poor solvents production was improved by ammonium sulphate precipitation and activated carbon treatment of the hydrolyzate. Addition of ferrons sulphate into the culture medium resulted in further solvents yield improvement to max. 33.5% in 60 h. As established already before, addition of butyric acid increased butanol accumulation. Unfortunately, the extent of pentose bioconversion into solvents was not assessed in this work which, considering the extent of the required pretreatment and treatment of the agro-wastes, has little practical value.

5 Current Trends in the Development of Acetone-Butanol Process

The acetone-butanol production from carbohydrates via an anaerobic microbial process was a dormant area for more than three decades [57]. The only operating large-scale plant in the Western hemisphere is a conventional process in South Africa [58]. Newly developing economic situation and skyrocketing prices of dwindling hydrocarbon feestocks have prompted reexamination of the old process which used to be the sole source of industrial solvents. The contemporary return to renewable resource materials is facilitated by recent breakthroughs in the genetic manipulation of microbial "catalysts" coupled with efficient utilization of the newly available electronic control and optimization techniques for improved biotechnology. The combination of these factors, together with economic and environmental pressures, makes biosynthesis production of certain compounds highly competitive (again) and as such it is receiving special attention. Production of fuel ethanol has been the focal point for some time. It does not take any special clairvoyance to see acetone-butanol next in line, butanol in particular having important uses as both a fuel and an industrial solvent. Its application in tertiary oil recovery make it an even more desirable compound. The Acetone-Butanol biosynthesis process, despite its well established tradition, or maybe just because of it, has to be put on a different highly advanced technological and scientific basis if it is to make a desired contribution in producing competitively priced industrial solvents. A number of improvements, all within reasonable reach, have to be considered as goals for current or intended effort in the novel process development. The R & D thrust in this area needs to focus on several facets of the process, namely:

5.1 The Microbial Catalyst

The key element in the process is the microorganism. Its performance in accumulating maximum 2.5% total solvents before shuting off its biosynthetic activity being inhibited by the accumulation of solvents (namely butanol, at ~1.3% level) is unsatisfactory. A microbe with increased tolerance to the solvents is essential. Elucidation of the inhibition mechanism is a high priority. "Engineering" the microorganism for higher alcohol tolerance and perhaps increased rates of solvent biosynthesis is the ultimate goal. This represents a major task involving demanding biochemical and even more demanding genetic manipulation and strain selection type of work. It is worth noting, that very little is known about the genetic makeup of strictly anaerobic microorganisms in general and solvent-producers in particular. A large amount of basic type of investigation is required in order to provide solid ground for any subsequent genetic manipulation. This type of work would be very desirable for another area of interest concerning.

5.2 The Metabolism of Unconventional Raw Materials — Pentoses, di- and oligo-Saccharides

Although the production *Clostridium* sp. are capable of utilizing pentose sugars, they do so at lower rates. The metabolic bottleneck(s) should be investigated and removed. Manipulation and/or optimization of the metabolic activities currently leading to excessive acid accumulation (or their slower reduction to corresponding alcohols) is very important. The property of some *Clostridia* of efficiently utilizing disaccharides (lactose, sucrose) and polysaccharides (starch, dextrin, xylans) can be enhanced and very effectively utilized in converting these raw or waste substrates into valuable solvent products [59,60].

5.3 The Development of Efficient Bioprocess Technology

A classical batch process is difficult to control and it cannot cope with the toxicity of the accumulating solvent products. It also does not utilize the microbial "catalyst" to the full extent, dumping it in the peak of its activity. The newly developed process should make the full use of the most productive "catalyst" by removing the accumulating solvents as they are being produced to prevent their toxic effect, and by retaining the biomass catalyst in the bioreactor for its most productive period of time. This approach can be based on operating a different type of a reactor with either physically or "hydraulically" immobilized cells. Several concepts offer themselves for investigation which should result in an unconventional bio-reactor design meeting the above-mentioned criteria. The novel bio-reactor should enable better process control and optimization of its performance resulting in improved process economics.

5.4 Product Recovery

To distill off the relatively low concentration of accumulated solvents (max. 2.5% total) is an increasingly more painful drawback of the process since the cost of heating the agueous solvent mixture is growing being coupled with the hydrocarbon price escalations. While the butanol boiling point of ~118 C poses a major problem, butanol immiscibility with water at higher concentrations partially offsets this disadvantage. A major breakthrough in the solvent recovery process would make the solvent fermentation manufacturing a single most attractive alternative. It could be based on advanced membrane technology, (selective) adsorption or absorption techniques by passing thus the high heating energy costs. Considerable research effort is being expended in this area in connection with fuel ethanol recovery and it is hoped that any advances in this area will have direct bearing on the acetone-butanol recovery technology as well.

5.5 Process Control and Optimization

Newly developed bioprocess technology based on continuous or semi-continuous flow systems, not necessarily continuous culture systems though, lends itself to application of increasingly more powerful, readily available and cheaper electronic control and microprocessor hardware. On-line process control and, indeed, optimization is a reality with an unprecedented potential. The software and algorithms are easily developed. The key elements for the control/optimization purposes, however, are measuring devices — sensors for various key process parameters of which concentrations (of reactants and products) are the most important ones. Fast and reliable analytical sensors are still expensive and represent a major, however, not unsurpassable bottleneck in the process control/on-line optimization scheme.

By understanding well the process kinetics and interdependence of individual parameters, which represents one of the main research objectives, predictions can be made as to the process behaviour and/or response to control parameter changes. This approach may prove useful in eliminating the need for some of the sensors, and in developing computerized process simulation schemes. Such schemes would assist in pinpointing the most important of the key process parameters for final operating adjustments as well as for experimental optimization of the process carried out in the laboratory. The simulation schemes may be of various degree of complexity, ranging from those for individual process operations to the simulation of the whole manufacturing process in its entirety. The importance of this approach should not be underestimated for the successful design and scale-up of the anaerobic microbial solvent production.

6 Conclusion

For development of a competitive novel concept of an acetone-butanol biosynthesis process it is imperative that *all* the above aspects be developed and applied simultaneously because of their high degree of interdependence. In a schematic acetone-butanol process flowchart on Fig. 4, the areas corresponding to the above points of (a) to (e) for further improvement and key R & D efforts are indicated.

It is most gratifying to see that relevant work is currently underway in several institutions [54,55,59,60,61,62], gaining momentum and successfuly addressing the most important facets of the challenging, exciting and important acetone-butanol biosynthesis process.

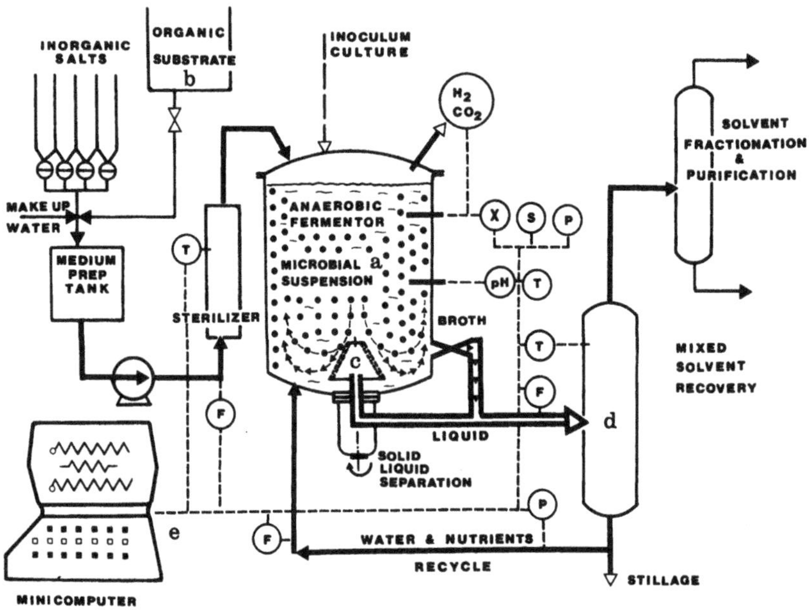

Fig. 4. Schematic flow chart of acetone-butanol production. The key process areas to be improved are indicated.
a) culture improvement by genetic manipulation;
b) rapid and efficient conversion of unconventional substrates;
c) use of contemporary bioprocess technology;
d) improved energy-saving solvent recovery techniques;
e) process control and on-line optimization

7 References

1. Aspinall, G. O.: Adv. Carbohydrate Chem. *14*, 429 (1959)
2. Peat, S., Thomas, G. W., Whelan, W. J.: J. Chem. Soc. 000, 456 (1952)
3. Brasch, D. J., Wise, L. E.: Tappi *39*, 581, 768 (1956)
4. Banerjee, S. K., Timell, T. E.: ibid. *43*, 489 (1960)

5. Schneider, H. et al.: Pentose fermentation by yeasts. in: Current Development in Yeast Research Adv. in Biotechnol., Stewart, G. G., Russell, I. (eds.), p. 81, Toronto: Pergamon Press 1981
6. Jeffries, T. W.: Biotechnol. Lett. *3* (5), 95 (1981)
7. Weckstrom, L., Leisola, L.: in: Adv. Biotechnol. 2, Moo-Young, M. (ed.), p. 21, Toronto: Pergamon Press 1981
8. Peppler, H. J.: in: Microbial Technology Peppler, H. J. (ed.), p. 145, New York: Reinhold Publishing Corp. 1967
9. Pasteur, L.: Extrait des proces-verbaux, Soc. Chim. Paris (Bull) 52 (1862)
10. Beijerinck, M. W.: Verhandel. Akad. Westenschappen Amsterdam, Afdeel. Natuurkunde, Sectie II, Deel I, (1893)
11. Duclaux, E.: Ann. Inst. Pasteur *9*, 811 (1895)
12. Emmerling, O.: Berinch. chem. Ges. *30*, 451 (1897)
13. Fitz, A.: ibid. *15*, 867 (1882)
14. Fitz, A.: ibid. *17*, 1188 (1884)
15. Grassberger, R., Schattenfroh, A.: Arch. Hyg. *42*, 219 (1902)
16. Grimbert, M.: Ann. Inst. Pasteur *7*, 353 (1893)
17. Gruber, M.: Zentr. Bakt. Infektionskrankh. *1*, 367 (1887)
18. Schardinger, F.: Zentr. Bakt. Parasitenk. Abt. II, *14*, 772 (1905)
19. Winogradsky, S.: ibid. Abt. II, *9*, 43 (1902)
20. Weizmann, C.: Brit. Pat., 4845 (1912)
21. Bergey's Manual of Determinative Bacteriology, Buchanan, R. E., Gibbons, N. E. (eds.), Baltimore: Williams & Wilkins 1974
22. Beesch, S. C.: Ind. Engng. Chem. *44*, 1677 (1952)
23. Brock, T. D.: Biology of Microorganisms, p. 144, Englewood Cliffs, N. J., Prentice Hall 1974
24. Fred, E. B., Peterson, W. H., Davenport, A. J.: J. Biol. Chem. *53*, 111 (1922)
25. Johnson, M. J., Peterson, W. H., Fred, E. B.: ibid. *91*, 570 (1931)
26. Racker, E.: Federation Proc. *7*, 180 (1948)
27. Lipmann, F.: Nature *138*, 588 (1936)
28. Dickens, F.: Biochem. J. *32*, 1626, 1645 (1938)
29. Rappaport, D. A., Barker, H. A., Assid, W. Z.: Arch. Biochem. Biophys. *31*, 326 (1951)
30. Lampen, J. O., Gest, H., Sowden, J. C.: J. Bact. *61*, 97 (1951)
31. Bolcato, V., Savola, M. E., Bettinetti, G. F.: Experientia *8*, 25 (1952)
32. Kaushal, R., Jowett, P., Walker, T. K.: Nature *167*, 949 (1951)
33. Dellweg, H.: Electron acceptors in anaerobic fermentations. in: Advances in Biotechnology *2*. Moo-Young, M. (ed.), p. 241, Toronto: Pergamon Press 1981
34. Valentine, R. C.: Bacteriol. Rev. *28*, 497 (1964)
35. Poston, J. M. et al.: J. Biol. Chem. *241*, 4209 (1966)
36. I. G. Farbenindustrie A. G.: Brit. Pat. 496428 (1938)
37. Schurmann, C., Vierling, K. (to I. G. Farbenindustrie A. G.): Germ. Pat. 635572 (1936)
38. Vierling, K. (to I. G. Farbenindustrie A. G.): Germ. Pat. 659389 (1938)
39. Tornescher Heffe, G.m.b.H. Ger. Pat. 920724 (1954)
40. Wiley, A. J. et al.: Ind. Engng. Chem. *33*, 606 (1941)
41. Müller, H. P., Rutishauser, M., Wiken, T.: Schweiz. Z. Allg. Pathol. u. Bacteriol. *20*, 517 (1957)
42. Schoedler, K.: Papier *12*, 645 (1958)
43. Sjolander, N. O., Langlykke, A. F., Peterson, W. H.: Ind. Engng. Chem. *30*, 1251 (1938)
44. Leonard, R. H., Peterson, W. H.: ibid *39*, 1443 (1947)
45. Langlykke, A. F., Van Lanen, J. M., Fraser, D. R.: ibid. *40*, 1716 (1948)
46. Kalnina, V. et al.: Latvijas PSR Zinatnu Akad. Vestis, Kim. Ser. *1961* (2) 269 (1961)
47. Underkofler, L. A., Fulmer, E. I., Rayman, M. M.: Ind. Engng. Chem. *29*, 1290 (1937)
48. Nakhmaninovich, B. M., Kameneva, L., Kalnina, V.: Latvijas PSR Zinatu Akad. Vestis *1963* (4) 120 (1963)
49. Compere, A. L., Griffith, W. L.: Evaluation of substrates for butanol production in: Proc. 35th Mtg. Soc. Ind. Microbiol. p. 509, Arlington, VA: Publ. Soc. Ind. Microbiol. 1979
50. Underkofler, L. A., Hunter, J. R. Jr.: Ind. Engng. Chem. *30*, 480 (1938)
51. Brown, P.: Bios (Mt. Vernon, Iowa) *32*, 77 (1961)

52. Nakhmaninovich, B. M., Yarovenko, V. L. Tr.: Vses. Nauch.-Issled. Inst. Prod. Brozheniya *1970* (19) 281 (1970)
53. Volesky, B., Szczesny, T., Neufeld, R.: Unpublished (1982)
54. Maddox, I. S.: Biotechnol. Letts. *4*, 23 (1982)
55. Soni, B. K., Das, K., Ghose, T. K.: Biotechnol. Letts. *4*, 19 (1982)
56. Hongo, M.: U.S. Pat. 2, 945, 786 (1960)
57. Hastings, J. J. H.: Acetone-Butyl Alcohol Fermentation in Econ. Microbiol. *2*, 31 (1978)
58. Spivey, M. J.: Process Biochem. *13*, (11) 2 (1978)
59. Lenz, T. G., Moreira, A. R.: Ind. Engng. Chem. Prod. Res. Dev. *19*, 478 (1980)
60. Volesky, B., Mulchandani, A., Williams, J.: Ann. N. Y. Acad. Sci. *369*, 205 (1981)
61. Leung, J. C. Y., Wang, D. I. C.: in: Proc. 2nd World Congr. Chem. Eng. *1*, Weber, M. E. (ed.), p. 348, Ottawa: Can. Soc. Chem. Eng. 1981
62. Haggstrom, L.: Immobilized cells of *Cl. acetobutylicum* for butanol production. in: Adv. Biotechnol. *2*, Moo-Young, M. (ed.), p. 79, Toronto: Pergamon Press 1981

Lignin: Biosynthesis, Application, and Biodegradation

H. Janshekar* and A. Fiechter
Dept. of Biotechnology, Swiss Federal Institute of Technology, ETH-Hönggerberg, CH-8093 Zürich, Switzerland

*New address: Petrogenetic AG, Limmatquai 112, CH-8023 Zürich, Switzerland

During the last few decades, research on lignin has focused on its biosynthesis, its possible applications, and its biodegradation. In order to better evaluate and understand the composition and possible uses of this material, this article reviews many of the most recent discoveries in lignin research and discusses several other aspects of lignin: its distribution and function in plants, the occurrence and sources of lignin, processes used to recover lignin, and its chemical structure. Since significant research is concentrated on the biological degradation, this aspect is also emphasized.

1 Introduction

Lignin is the characteristic cementing constituent between the cell walls of woody tissues. The word lignin is derived from the Latin word 'lignum,' meaning wood. It does not represent a definite, uniform compound, but is a collective form for substances that have very similar chemical properties but very different molecular weights. The molecular weight of lignins may reach the range of 100,000 daltons or even greater. A considerable part of the photosynthetic activity in plants is devoted to the conversion of atmospheric carbon dioxide to lignin. Lignin constitutes about 40% of the solar energy stored in plants. Hence it plays a highly significant role in the carbon cycle.

Questions concerning the existence and nature of lignin date back to before the First World War. These discussions were, however, in connection with the chemistry of pulping and the chemistry of natural tannins. No direct attempt was undertaken to elucidate the biosynthesis and biochemistry of lignification until several decades ago. It was not clear whether lignin was a refuse product of nature nor whether it belonged to the phenols. For almost thirty years, progress made on the lignin problem was depressingly slow. Thirty-five years ago, lignin was still considered to be an artefact from the condensation of phenols and sugars during the working-up processes of wood. The development of lignin chemistry was closely connected with the question of whether or not lignin is an aromatic substance. Three decades ago, when it became increasingly evident that ligning was, in fact, of phenylpropanoid origin, investigations were focused on finding how these precursors are formed in nature and how they are transformed into lignin.

In the 1960s, many scientists, particularly Freudenberg [69], Brown, Higuchi [95-99], and Sarkanen [200, 201] and their associates, made significant contributions to progress in this complicated subject (for citations, see Gross [78] 1980). Considering the limitations of analytical chemistry in those years, it becomes clear that the progress would have been very difficult. 'Had it not been for the confidence in an orderly constitution of polymers that I had received from my early work on cellulose, I should never have risked attacking the problem of lignin, for not even the haziest notions about the polymeric molecules were developed at that time' Freudenberg, 1968 (see Ref. [78]).

A considerable breakthrough came about when it was found that lignin represents a complex macromolecule originating from the random oxidative polymerization of hydroxylated cinnamyl alcohols. From that time on, intensive efforts in lignin biochemistry research have brought us substantial knowledge about the constitution of lignin. The understanding of the composition and structure of lignin is now so detailed that, in most cases, logical correlations between the chemistry of lignin and the biochemical events involved in its synthesis can be found.

Research on the biological degradation of lignin started even later than that on lignin biosynthesis. Lack of appropriate understanding of the chemistry of lignin until the 1960s and the lack of strong interest and research support were the major reasons for this delay. Research on the biodegradation of lignin was mainly activated by the work of Kirk [119–121] at the U.S. Forest Products Laboratory in Madison, Wisconsin, in the early 1970s. Kirk and his colleagues have published over 60 scientific works in this area.

Hope of the eventual application of lignin, this most abundant renewable material, has stimulated many other research groups in the world in the last few years to contribute to this complicated subject. The importance of the subject led to the organization of an international seminar in 1978. Kirk, Higuchi, and Chang [121] edited the proceedings of this seminar, which was published in 1980. This two-volume book covered most activities made in the area of lignin biodegration up to that time. Crawford [43], in a comprehensive book, reviewed the major progress made in lignin biodegradation and transformation research up to 1981. However, the number of papers published on this subject has increased so intensively in the last two years that the need for a new review has become necessary.

Attempts continue for finding new lignin-degrading organisms, improving methods for the study of lignin degradation, elucidating the mechanisms of lignin degradation and improving the the rates of biodegradation. The present review documents the highlights of existing knowledge of this topic, and provides a view of the current research and activities, with emphasis placed on the biological degradation of lignin.

2 Biosynthesis

The photosynthetic assimilation of atmospheric carbon dioxide in plants leads to the formation of carbohydrates. Carbohydrates are metabolized via a shikimic acid pathway and converted to phenylpropane amino acids (Fig. 1). These amino acids supply precursors for the synthesis of plant proteins, lignin, and flavonoids.

The first step in the synthesis of the above mentioned precursors is the *trans*-elimination of ammonia from L-phenylalanine to form *trans*-cinnamic acid. Successive hydroxylation and methylation reactions convert cinnamic acid to several substituted acids. The distribution and proportion of these acids in various vascular plants are different [96]. The acids then enter biosynthetic pathways leading to lignin, flavonoids, stilbenes and other compounds.

In the case of lignin biosynthesis, cinnamoyl-CoA esters of substituted cinnamic acids are reduced to cinnamyl alcohols or monolignols, i.e., *p*-coumaryl, coniferyl, and sinapyl alcohols. Once the cinnamyl alcohols are formed, their hydroxyl group is oxidized by peroxidase or laccase, thus yielding free radical species in mesomeric forms (Fig. 2). The required hydrogen peroxide is formed in a complex reaction requiring NAD(P)H and Mn^{2+} [78]. These radicals then couple in a non-enzymatic and random fashion to form dilignols, oligomeric intermediates, and finally lignin macromolecules. Depending on its structural monomeric units, the synthesized lignin could be gymnosperm (dehydrogenation polymers of coniferyl alcohols), angiosperm (mixed dehydrogenation polymers of coniferyl and sinapyl alcohols), or grass-type lignin (mixed polymers of coniferyl, sinapyl, and *p*-coumaryl alcohols [95]).

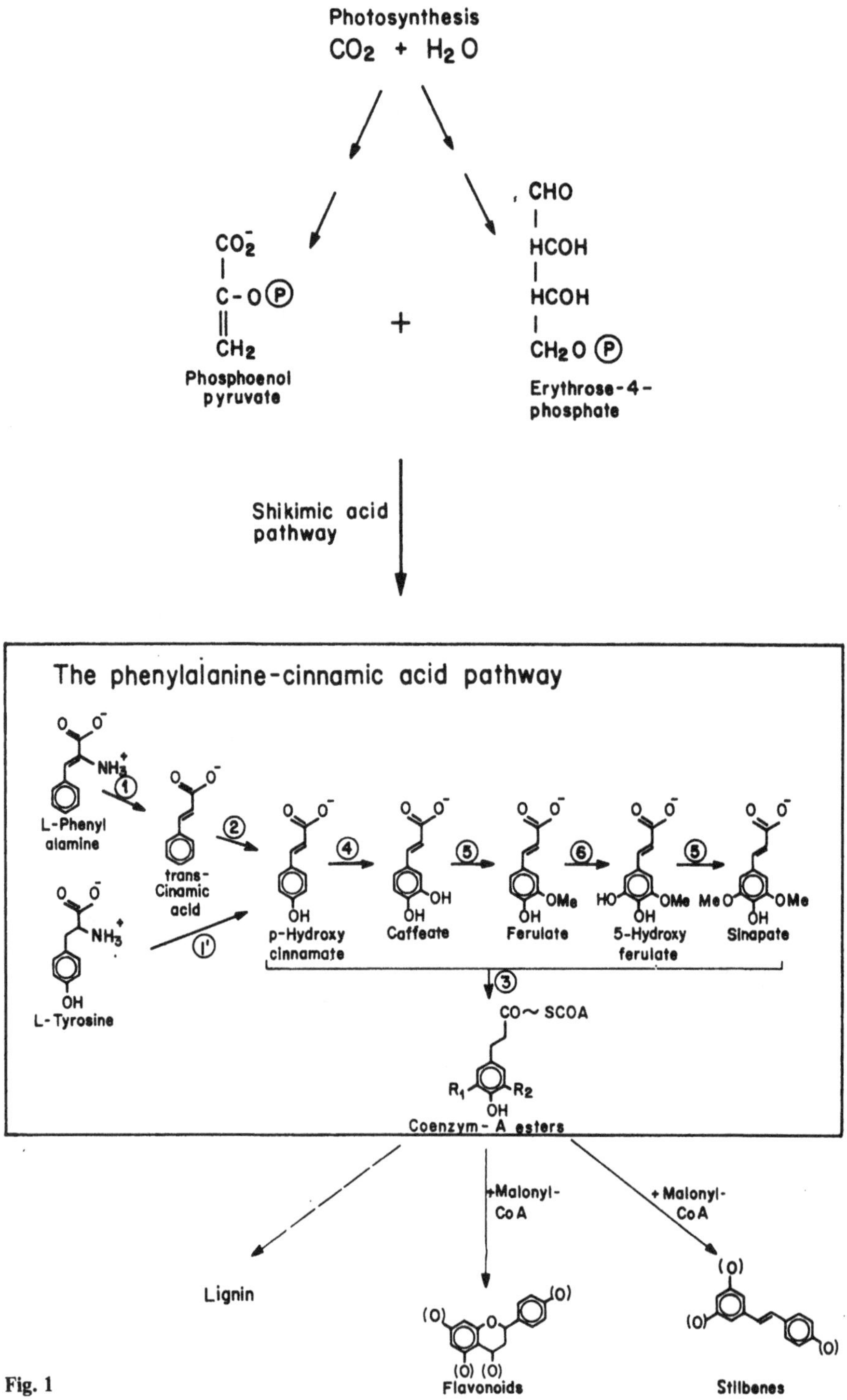
Photosynthesis
CO2 + H2O
Phosphoenol pyruvate
Erythrose-4-phosphate
Shikimic acid pathway
The phenylalanine-cinnamic acid pathway
L-Phenyl alamine
trans-Cinamic acid
L-Tyrosine
p-Hydroxy cinnamate
Caffeate
Ferulate
5-Hydroxy ferulate
Sinapate
Coenzym-A esters
+Malonyl-CoA
Lignin
Flavonoids
Stilbenes

Fig. 1

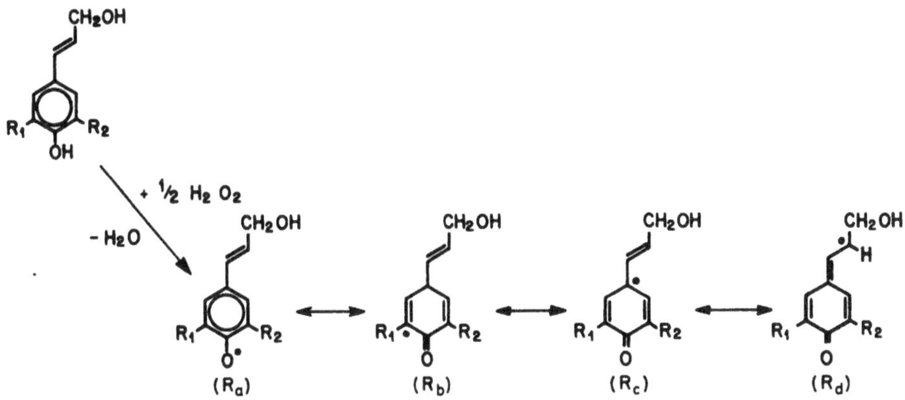

Fig. 2. Formation of phenoxy radicals, the immediate precursors for chemical polymerizations of lignin in woody plants, by peroxidase. $R_1 = R_2 = H$: *P*-coumaryl alcohol; $R_1 = H$, $R_2 = OCH_3$: coniferyl-alcohol; $R_1 = R_2 = OCH_3$: sinapyl alcohol [78]

It is still not clear whether any plant organelles are involved in the transport of lignin monomers to the cell walls.

Most of the work done to elucidate the biosynthetic pathway of lignin has consisted of feeding experiments using labeled precursors or enzymic synthesis of lignin precursors. Not much is known about the in vivo lignification. Plant cell cultures have proved to be an excellent tool for research in this field [83]. To date, the factors that play a regulatory role in biosynthesis of lignin are the supply of lignin precursors, the transport of lignin monomers into the cell wall, and the polymerization of cinnamyl alcohols. The individual significance of each of these factors, however, remains to be clarified.

o-methyltransferase (OMT) and *4-coumarate*: CoA ligase are among the most studied enzymes for the synthesis of lignin precursors [128, 131, 143]. The only uncharacterized enzyme in lignin biosynthesis is ferulate-5-hydroxylase, which is a key enzyme for separating lignin biosynthesis into guaiacyl- and syringyl-lignin pathways. The formation of guaiacyl lignin rather than syringyl lignin in gymnosperms is considered to be due to the absence of ferulate-5-hydroxylase, the poor affinity of OMT towards 5-hydroxy ferulate, and the lack of activation and/or reduction of sinapate [95].

The two enzymes involved in the reduction of cinnamyl-CoA esters have been identified as cinnamoyl-CoA reductase and cinnamyl alcohol: $NADP^+$ dehydrogenase. Reductase and dehydrogenase from angiosperm (soybean) and gymnosperm (spruce) show pronounced differences in substrate specificity [142]. The regulatory role of cinnamyl alcohol dehydrogenase in the formation of lignin precursors has been

◀ **Fig. 1.** Reaction sequence of phenylpropanoid metabolism [77]. Enzymes: ① phenylanaline ammonia-lyase (PAL). ①' Tyrosine ammonia-lyase (TAL), only in grasses. ② Cinnamate 4-hydroxylase. ③ 4-Coumarate: CoA ligase or 4-Coumaroyl-CoA synthase. ④ 4-Coumarate 3-monooxygenase. ⑤ S-adenosylmethionine (SAM): caffeate 3-0-methyltransferase (OMT). ⑥ "Ferulate 5-hydroxylase" (hypothetical)

studied and reported [96, 130]. According to Schmid and Grisebach [204], when intensive synthesis of coniferyl alcohol takes place, coniferin, a storage product, acts as a reservoir to augment the supply of precursors for lignifying cells. An enzyme (glucosyltransferase) that may play a role in the formation of coniferin has also been isolated.

3 Distribution and Function in Plants

The spaces between cellulose fibrils in woody vascular tissues are filled with lignin and hemicellulose. Lignin is most concentrated in and between the primary walls (middle lamella) of adjacent vascular plant cells [8]. The concentration of lignin is lower in the fiber wall than in the middle lamella. However, since the fiber wall occupies about 90% of the volume of the cell tissue, the largest portion of plant lignin is located in this part.

Lignin performs a number of functions essential to the life of plants. A summary of how lignin properties affect plant characteristics is in Table 1. Lignin functions

Table 1. Contribution of lignin in plant properties

Function of lignin	Plant property
Energy storage system (potential and entropic)	Photosynthesis
Permanent bonding agent between cells, resulting in composite structure of wood	Resistance to mechanical stresses such as compression and blending
Impediment to penetration of destructive enzymes, inhibitor to enzymic degradation of other plant components	Resistance to biochemical stresses such as microbial attack, infection and wounding
Antioxidant, uv light stabilizer, flame retardant	Resistance to chemical stresses such as atmospheric degradation, uv light radiation
Water proofing agent	Resistance to physical-chemical stresses, response to humidity, water balance and transport, nutrient transport

as a binding and encrusting material for the cell wall constituents, and gives rigidity to the plant cell wall which helps the cell resist external conditions. Lignin also decreases water permeation across cell walls in the conducting xylem tissues, thus preventing water leakage from cell walls. The water-permeation-reducing property of lignin also plays an important role in the internal transport of water, nutrients, and metabolites in the plant.

Lignification is a common response of the plant to infection or injury [77, 126]. The healing process of injured plants involves lignification and the suberization of surface cells, which is associated with an increased resistance to infection [71]. Fungal chitin may be one of the elicitors of the lignification response to injury [177].

4 Occurrence and Sources of Lignin

Lignocellulosic materials comprise about 95% of the earth's land-based biomass, about 25% of which is lignin. Lignin-like polymers are incorporated into the macromolecules of lignite and bituminous coals [90]. Sources of lignin are shown in Table 2; the composition of some lignocellulosic materials is presented in Table 3.

Table 2. Sources of lignin

Native lignocellulose	Process liquor streams	Solid wastes
Wood, cotton, vegetable fiber crops, cereal straws, corn stalks	Kraft spent liquor, sulfite spent liquor, pulp mill waste streams	Municipal and urban, cattle manure, bark and foliage from pulp industry

Table 3. Composition of some lignocellulosic materials [68]

	Approximate composition (% dry weight)		
	Cellusose	Hemicellulose	Lignin
Coniferous wood	40–50	20–30	25–35
Deciduous wood	40–50	30–40	15–20
Cotton	94	2	0
Bagasse	40	30	20
Nut shells	25–30	25–30	30–40
Corn cobs	45	35	15
Corn stalks	35	25	35
Wheat straw	30	50	15
Paper	85–99	0	0–15
Newsprint	50	20	30
Sorted refuse	60	20	20

Lignin also occurs in the waste materials from several industries such as forestry, agriculture, paper-making, and lumbering. The wood-carbonizing waste of the textile industry also contains acidic ligneous wastes. Together, all of these industries, which usually operate at large capacities, yearly dispose astronomical quantities of lignocellulosic wastes. For example, nearly one half of the original wood used in pulp manufacture is converted to sulfite waste liquor, with 70—80% of its organic material consisting of lignin. Lignin is also a high-volume by-product of most of the currently envisioned biomass-processing schemes such as chemical- and enzymatic-biomass saccharification.

There is no exact estimate of the world-wide occurrence of lignin-containing materials. According to various sources, the annual production of lignin wastes in the U.S. should be between 900 to 3000 million tons. In 1975, 300 million tons of forest products were consumed in the U.S., which corresponds to the production of 70 million tons of lignin [62]. U.S. Kraft mills produce about 21 million tons per year of Kraft lignin [6]; about three times that amount should be the annual production of the world.

A part of this lignin is used as a source of processing energy such as in the Kraft pulping process. However, considerable amounts of this material are disposed in various ways. For example, large amounts of lignin sulfonates are still being released into natural waters. It is estimated that 0.16 million tons of this material were released into the Gulf of Bothria from Swedish sulfite mills in 1975, or 13% (W/W) of the pulp produced that year [197].

5 Processing of Lignocellulosic Materials for Lignin Recovery

There are two categories of processes for the chemical separation of the three main components of lignocellulosic biomass:

(1) processes that dissolve and recover lignin and

(2) processes that remove holocellulose.

Detailed explanations of these processes are beyond the scope of this review article; however, the literature in this area has been reviewed by Allen et al. [6] [p. 22]. No matter which type of processing is used, physical pretreatment such as milling, radiation, or heating can increase the separation efficiency and facilitate the release of one or the other lignocellulosic component. In the following sections, the main chemical and thermal methods used for the separation of the three main components of lignocellulosic biomass are described briefly.

5.1 Hydrothermal Processes

In this process, lignocellulosic materials are hydrolyzed in acidic conditions at relatively high temperatures (170–280 C) [201]. The end products of the hydrothermal process are cellulosic fibers, lignins, and aqueous solutions of sugars, predominantly xylose. Physical factors, such as average pore size and volumetric swelling of chipped or chopped raw materials, can affect the extent of thermal hydrolysis.

Hydrothermal treatment causes lignin to undergo several structural changes. The internal bonds in lignin such as α-alkyl and β-aryl ether types are cleaved. Further, partial hydrolysis and breakage of lignin-carbohydrate bonds, as well as condensation reactions between lignin moieties and between lignin and other hydrolysis products such as furfural or hydroxymethyl furfural take place. Lignin obtained using typical lignocellulosic technologies such as dilute and concentrated acid hydrolysis has a poor chemical reactivity. Selke et al. [209] have produced evidence that HF saccharification may be superior in respect to the reactivity of the produced lignin, the yield of glucose, and the chemicals consumed. Hydrothermal processes can be divided into two main categories: aqueous hydrolysis extraction and organosolv delignification.

Aqueous hydrolysis extraction: Hydrolysis may be done using water or steam, which is then followed by defibration and the recovery of lignin, usually using methanol. A process uses a steam explosion, followed by enzymatic hydrolysis to produce pure lignin [18]. The steam explosion pretreatment preserves the high quality of the lignin and prevents repolymerization during the hydrolysis process. In this process, over 200 kg of lignin can be produced per ton of wood chips at a cost of 44 cents per kg (1981 prices).

Organosolv delignification: In this process, hydrolysis and extraction of lignin is performed simultaneously. The solvents usually used are mixtu.. es of lower alcohols such as ethanol, n-butanol, and methanol with water. Ketones (acetone), ethers (dioxane), phenols, and triethyleneglycol may also be used. Reactions may be carried out with or without catalysts (acidic or basic) or may also be buffered.

Comparisons made between the two hydrothermal delignification methods have shown that the organosolv method generally achieves more selective and extensive delignification than the steam hydrolysis extraction process [201]. This is thought to be due to the suppression of hemicellulose hydrolysis by organic solvents and to the dissolution of lignin which takes place during its hydrolysis, consequently resulting in less lignin condensation [139].

5.2 Alkaline Processes

In the alkaline process, lignocellulosic material is treated with sodium hydroxide or ammonia. The alkaline pretreatments are applied to a variety of lignocellulosic substrates, primarily to increase their digestibility for feeding to ruminants. This treatment is also used in alkaline pulping processes (soda and Kraft). In the soda pulping process, wood is treated with hot sodium hydroxide solution under pressure. The Kraft or sulfate pulping process uses a $NaOH—Na_2S$ mixture (white liquor) to dissolve lignin from wood carbohydrates.

Delignification can also be facilitated by the addition of anthraquinone (AQ) to pulping liquors. The anthrahydroquinone (AHQ), reduced from AQ by carbohydrates, interferes with alkaline reactions of the lignin and accelerates its degradation [150]. Comparisons of sulfide and anthraquinone catalysts to alkaline pulping has led to the realization that the mechanisms of delignification are not the same for both catalysts [229]. Delignification can also be enhanced by the addition of sodium perborate, hydrogen peroxide, and oxygen [70] or by pretreatment using NO_2, O_2 [198] or acid prehydrolysis [123]. Primary aliphatic amines and diamines can also accelerate delignification. With an amine, the rate of β-O-4 ether cleavage of blocked phenolic units can be increased without increasing alkali concentration [172]. It has been proposed that the role of amines is of chemical rather than physical nature [1].

6 Chemical Structure

6.1 Native Lignin

Lignin is composed of highly branched polymeric molecules consisting of phenylpropane based monomeric units linked together by different types of bonds, including alkyl-aryl, alkyl-alkyl, and aryl-aryl ether bonds. The relative proportion of the three cinnamyl alcohol precursors (Fig. 2) incorporated into lignin varies not only with the plant species, but also with the plant tissues and the location of the lignin within the plant cell wall. Ecological factors such as age of the wood, climate of the environment, plant sustenance, amount of sunlight, etc. also affect the chemical structure of lignin.

A major problem in studying the chemistry of lignin has been the difficulty in isolating lignin from plant materials without secondary reactions. It is generally accepted that protolignin cannot be solubilized without depolymerization and degradation of the macromolecular network of the cell wall. Cleavage of certain bonds, changes in the extent of lignin condensation, changes in the content of certain functional groups, and alteration of chemical and physical properties of lignin are likely to take place during separation of lignin from woody tissue. Therefore, the method of isolation used for separating lignin from plant tissues can cause significant alterations in the lignin structure. Amer and Drew [8)] have summarized some of the changes in lignin characteristics that are caused by various isolation techniques. When one considers these drawbacks, it is understandable why the elucidation of the structure of lignin has been most difficult in the chemistry of natural polymers and why it is difficult to find a model that can generally represent the lignin structure.

Various methods or combinations of methods have been used for the illustration of lignin structure. These include enzymatic dehydrogenation of *p*-hydroxycinnamyl alcohols, structural determination of lignin degradation products, spectral and functional analysis, and structural simulation of lignin by computer [160)].

Based on the degradation products from softwood and hardwood lignin obtained by hydrolysis with dioxane-water and by catalytic hydrogenolysis, Sakakibara [196)] has proposed a tentative structural model of 28 units for softwood lignin. Despite the small number of units, analytical comparisons of the suggested lignin model with softwood (spruce) milled-wood lignin (MWL) showed a good accordance (Table 4).

Some of the interlignol bonds and lignin's functional groups are shown in Tables 5 and 6. Bulky side groups of lignin such as methoxyl functions have been found to expand intermolecular distances, whereas hydroxyl groups reduce their mean value [89)]. The most important linkages in the lignin structure are β-O-4, 5-5 biphenyl, β-5, and

Table 4. Analytical comparisons of spruce-MWLs and lignin structural models [196)]

Lignin and models	C_9-Formula	Methoxyl-free formula	Degree of dehydro-genation	Moles of added H_2O
Coniferyl alcohol	$C_9H_9O(OMe)$	$C_9H_{10}O_2$		
MWL (Björkman 1957)	$C_9H_{8.83}O_{2.37}(OMe)_{0.96}$	$C_9H_{9.05}O_2(H_2O)_{0.37}$	0.95	0.37
MWL (Freudenberg 1968)	$C_9H_{7.95}O_{2.40}(OMe)_{0.92}$	$C_9H_{8.07}O_2(H_2O)_{0.40}$	1.93	0.40
Model A	$C_9H_{7.93}O_{2.39}(OMe)_{0.93}$	$C_9H_{8.08}O_2(H_2O)_{0.39}$	1.92	0.39
Model B (alternative)	$C_9H_{7.96}O_{2.43}(OMe)_{0.93}$	$C_9H_{8.03}O_2(H_2O)_{0.43}$	1.97	0.43
Mean value of A and B	$C_9H_{7.95}O_{2.41}(OMe)_{0.93}$	$C_9H_{8.06}O_2(H_2O)_{0.41}$	1.94	0.41

Model A: $C_{252}H_{222}O_{67}(OMe)_{26}$; M.W. 5124
Model B: $C_{252}H_{223}O_{68}(OMe)_{26}$; M.W. 5141

Table 5. Major internal linkages in lignin [119,203,201]

Type of linkage	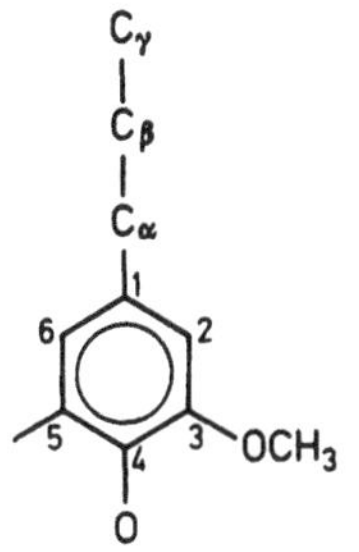	% of total phenylpropane units	
		Softwood (spruce)	Hardwood (birch)
β-Aryl ether (β-0-4)		45–48	60
α-Aryl ether (α-0-4)		6–8	6–8
Diphenyl ether (4-0-5)		3.5–8	6.5
α-Alkyl ether (α-0-γ)		Small	Small
Biphenyl (5-5′)		9.5–17	4.5
β-1		7–10	8
β-5		14	
β-β		3	
Phenylcoumaran structure		9–12	6
Lignin-carbohydrate (α-0-R)		Unknown	Unknown

Table 6. Major functional groups of lignin [203]

Functional group	% of total softwood phenylpropane units
Aliphatic OH	100
Phenolic OH	25
Methoxyl	93
Carbonyl	18
End groups	9
Uncondensed guaiacyl groups	45

β-1. The arylglycerol-β-aryl ether bond is the most prevalent linkage in lignin. A small percent of lignin units are involved in pinoresinol structures [145]. Aliphatic double bonds are few. The frequency of α-alkyl and γ-aryl ether bonds has not yet been established. Studies of the self-condensation reactions of lignin model compounds vanillyl and veratryl alcohol have suggested that benzyl ether and diphenylmethane structures may be formed within lignin as it ages in the plant cell wall [93]. Sudo and Pepper [213] have recently isolated a dimer with a benzyliso-chrome structure from the hydrogenolysis products of aspen lignin. They concluded that such structures with C_α and C_β linkages should be present in wood lignin.

The occurrence of non-cyclic benzyl aryl (α-O-4) ethers in lignin was first proposed by Freudenberg and Friedmann [69] and was doubted by Leary [135,136]. Nimz [167] has provided evidence in favor of the non-cyclic α-O-4 bonds in lignin. Lundquist [146] has reported that the MWLs from spruce and birch contain only a few units with non-cyclic benzyl aryl ether linkages.

Environmental chemicals such as pesticides can be incorporated into lignin in a biochemically defined system. Chlorinated anilines, the components and primary metabolites of a number of widely-used pesticides and other agrochemicals, have been incorporated covaletly into lignin [219]. NMR studies and experiments using catalase treatment suggest that a major mechanism of co-polymerization consists of the nucleophilic addition of anilines to the benzylic α-carbon of lignol quinonemethide intermediates of lignin precursors. Since anilines seem to be incorporated into lignin by a simple nucleophilic addition, the possible use of a wide variety of chemicals such as NH_2, OH, SH, or COOH substitutes in lignin cannot be excluded [219].

In terms of physical properties, lignin is an amorphous polymer that has no crystallinity. The mode of polymerization during lignin biosynthesis makes lignin optically inactive. The amorphous nature of lignin has been studied using various techniques such as broad-line nuclear magnetic resonance, differential scanning calorimetry, viscoelasticity, and x-ray diffractometry [89].

The molecular weight of lignin produced from various biomass sources (softwood and hardwood, bagasse, straw) and by different processes varies extensively. Lignin may be oxidized in air. It is insoluble in water, difficult to be wetted and difficult for microorganisms to penetrate. Lignin is generally acid stable but can be solubilized under alkaline conditions. The study of the coagulation behavior has shown that lignin interferes with retention acids, internal sizing, etc., which makes its removal from waste water difficult [137].

6.2 Lignin-Carbohydrate Complex (LCC)

Lignin is closely bound to cellulose and hemicellulose in plant cell walls. The separation of lignin from carbohydrates in LCC cannot be achieved by using conventional analytical methods such as gel filtration, electrophoresis, ultracentrifugation, and hydrophobic-interaction chromatography.

Azuma et al. [13] have suggested a method for the isolation of LCC from MWL. Lundquist et al. [149] were able to prepare MWL with a low carbohydrate content from birch using liquid-liquid extraction. The LCC prepared from MWL of *Pinus densiflora* was made up of three fractions with apparent weight-average molecular weights of 5.0×10^3, 5.0×10^5, and small portions (approximately 4%) of molecules as large as Blue Dextran [13].

The chemical bonds between lignin and carbohydrates have been studied [46, 110, 171] Sequential enzymatic and chemical hydrolysis of LCC from aspen wood gave evidence for arabinan type polysaccharides and lignin-saccharide ether bonds in the complex. Milled-wood enzyme lignin, the unfractioned lignin residue obtained by polysaccharidase digestion of vibratory ball-milled hardwood and softwood, was shown to have similar L-C bond frequencies [171]. Based on the analytical results of the reaction products of the quinonemethide of guaiacylglycerol-β-guaiacyl ether with sugars and dehydrogenative polymerizate (DHP)-polysaccharide complexes, Higuchi [96] has postulated that LCCs are formed via quinonemethide intermediate reactions that take place in the plant cell walls during lignification.

6.3 Lignin-Protein Complex

Lignin preparations of woody material contain, besides carbohydrates, significant amounts of protein. The Klason lignin preparation from the cell wall of *Pinus elliottii*, for example, contains 9% protein [231]. Evidence indicates that lignin makes covalent bonds with cell-wall protein and ammonium-N of the nitrogen-containing compounds in soil [225, 226]. Amino acid profiles of lignin preparations have suggested that the polymerizing lignin links covalently with the cell wall glycoprotein and that these bonds may be formed preferentially with hydroxyproline [231].

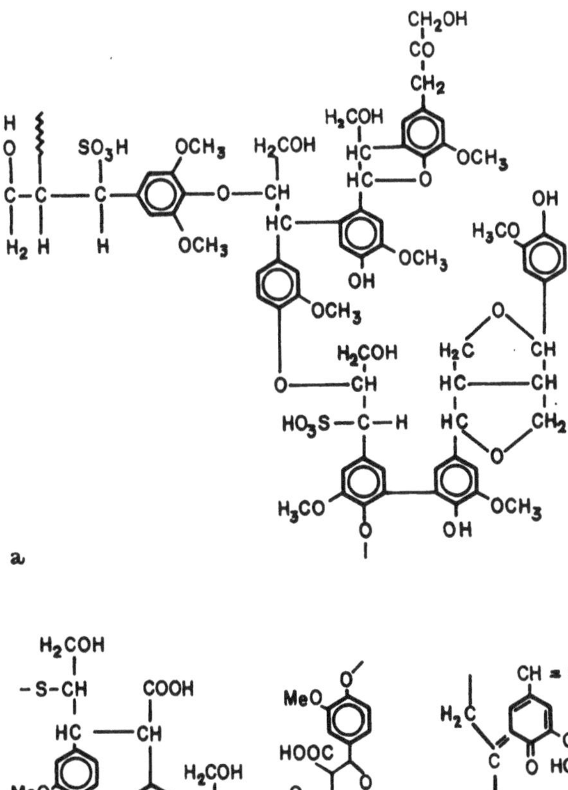

a

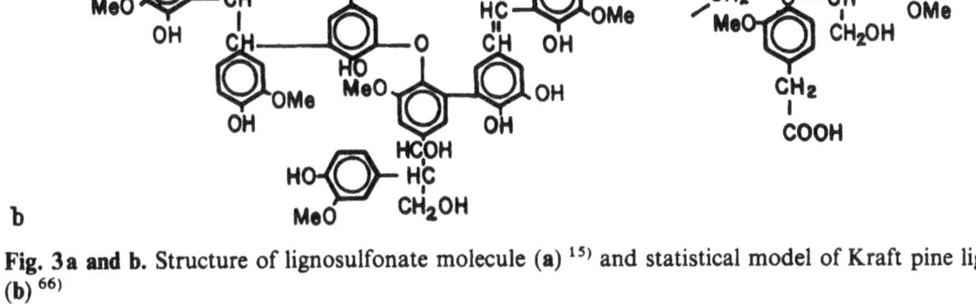

b

Fig. 3a and b. Structure of lignosulfonate molecule (**a**) [15] and statistical model of Kraft pine lignin (**b**) [66]

6.4 Industrial Lignins

Industrial or modified lignins are mainly by-products of sulfite and sulfate wood pulping operations. Industrial lignins, because of the extreme processing conditions that they pass through, have a different structure from that of native lignins. The change in lignin structure is dependent on the type and degree of treatment applied to the lignocellulosic material (Fig. 3). The general organic structures of industrial lignins, however, have not yet been completely determined.

Structural studies of lignin based on pulping aspects started with softwood lignin and continued with lignins from hardwood and grass. During sulfite reactions, sulfonation occurs mostly in the α-position of the side chain (Fig. 3a). In the Kraft pulping process, sulfur incorporates into the lignin structure during the dissolution process. Sulfur incorporated into the Kraft lignin has been shown to bond at the β-position in the phenyl-propane side chains (Fig. 3b). Participation of sulfoxide and sulfone linkages, as well as thiol and polysulfide functions in Kraft lignins, have also been reported [66]. Antraquinone additions during alkaline pulping processing cause the aryl ether bonds in phenolic β-O-4 structural units to cleave [27]. Kern [117] has listed the main differences between industrial lignins that arise from the use of acidic sulfite or from the Kraft pulping process (Table 7). The chemistry of delignification has been reviewed by Gierer [73–75]. Kinetics of delignification has been studied by Yan [234]. He has developed a mathematical model to identify the cleavable linkages and predict the changes of molecular weight of lignin on delignification. The model has been verified by experimental data obtained from isothermal delignification. For reactions of lignin with alkaline hydrogen peroxide, see [72].

The treatment of lignocellulose by dioxane, aceton, dimethyl-sulfoxide, and acetic acid in warm acid solutions open to the air can cause an increase of condensed structures, aliphatic double bonds in phenyl cumarones and stilbenoides, α, β-keto

Table 7. Main chemical reactions in pulping processes [117]

Kraft pulping	Acidic sulfite pulping
1. Aryl-alkyl ether cleavage: a) Cleavage of α- and β-ethers b) Liberation of new phenolic units c) Limited demethylation of methoxyl groups creates catechol moieties	1. Aryl-alkyl ether cleavage: a) α-Aryl ether linkages are opened by hydrolysis, formation of the reactive benzylium ion b) β-Aryl ether linkages (phenolic and non-phenolic) resist sulfite treatment
2. Strong modifications of the side chains: Formation of intermediate episulfides and epoxides, partial elimination of γ-C group as formaldehyde	2. Introduction of sulfonic acid goups mainly in α-position of the side chain
3. Condensation reactions compete with depolymerization caused by ether cleavages	3. Condensation reactions, formation of new C-C bonds
Properties: Soluble as salt Insoluble as acid	Properties: Soluble as acid and as salt

groups, phenolic hydroxyl and free radicals and the decrease of methoxy and aldehyde groups [227]. The condensation of lignin occurs during acid hydrolysis of wood with concentrated hydrochloric acid and at moderate temperatures. Condensation increases with increasing exposure time and with increasing temperature and acid concentrations [176]. Autohydrolysis (steaming) of wood also causes lignin to become more condensed in terms of new carbon-carbon bonds [223]. Increasing autohydrolysis time causes the guaiacyl-syringyl type lignin to change to syringyl-deficient type lignin [34]. Further understanding of the nature of the changes that occur in lignin during autohydrolysis of wood can be obtained from the work reported by Lora and Wayman [140].

7 Applications

Efforts to develop commercial applications for lignin have concentrated on the use of lignosulfonates and to a lesser extent on the use of Kraft lignin. Possible applications of lignin are listed in Table 8. In the pulp industry, the most common use of

Table 8. Alternative uses for lignin

Energy		
Polymers and modified polymers		
Filters, rubber reinforcment, carriers for controlled release in fertilizers and pesticides, dispersants, emulsion stabilizers, complexing agents, precipitants, coagulants, stabilizer applications e.g. as antioxidants, ion exchange		
Prepolymers		
Polyphenolics, resins and extenders, foams, adhesives		
Fragmentation and chemical conversion		
Hydrogenation	→	phenols, hydrocarbons, substituted mononuclear phenols
Hydrolysis	→	phenols, catechols, substituted phenols
Oxidation	→	vanillin, dimethyl sulfide, methyl mercaptan
Alkali fission	→	phenolic acids, catechols
Pyrolysis	→	acetic acid, phenols, substituted mononuclear phenols, CO, CO_2, CH_4, and H
Fast thermolysis	→	acetylene, ethylene

lignin is burning as fuel. Lignin's high energy content and high fuel value, as well as its high separation costs, structural complexity and variability, low reactivity and by-product status are reasons why lignin has not yet been used for other applications. Finally and most important, lignin products have always had to compete with similar products produced from petroleum. However, with increasing petroleum prices, it is only a matter of time before lignin-derived products become economically attractive.

7.1 Energy

Lignin has a high calorific value (7.1 kcal g^{-1}) compared with that of coal and oil. It comprises nearly 40% of the fuel value of softwood, although constituting only

25–35 % of its dry mass [52] (Table 3). Because of its relatively high calorific value, most lignin from pulping processes is used as fuel in chemical recovery processes of pulp plants. Dried hydrolytic lignins, along with wood bark and other wood wastes, are used as a combustion fuel in the furnaces of steam boilers. Therefore, the alternative uses of lignin is presently governed by the economics of the fuel supply.

7.2 Macromolecular Application

The macromolecule of lignin has several physical properties and therefore can be used as a polymer in various systems directly or after certain chemical modifications. Lignin is adsorbed onto suspended particles, imparting a negative charge that causes the particles to repel each other through an electrokinetic effect. This may hinder suspended particles from directly contacting each other or the surrounding aqueous medium through the film barrier effect. The suspended solids have different suspension tendencies, depending on their size and density, and thereby can be separated. Lignin also helps prevent crystalline growth in aqueous solutions and hinders the formation of insolubles. Lignin can reduce the viscosity of slurries and allow an increased concentration of solids without a rise in viscosity. Lignin-stabilized emulsions are highly resistant to breaking, for example, from pH or temperature variations or from the action of electrolytes.

Lignosulfonate-based products are now used in dye systems, agricultural chemicals, pesticide formulations, waste-water treatment, industrial cleaners, sequestrants, conditioners for oil-well drilling and cementing systems, asphalt emulsions, gypsum stucco slurries, clay deflocculation and conditioners, mineral beneficiation processes, thickeners, gelling agents, and for many other purposes.

One potential application for lignin is in the oil industry. Lignosulfonate thinners or deflocculants are the most versatile and important chemicals used in water-based drilling fluids. Lignosulfonates, particularly at high concentrations, control filtration and inhibit disintegration and disperions of shale cuttings. Lignosulfonates, because of their excellent surfactant and emulsion-stabilizing properties, can be used in place of commercial petroleum sulfonates in petroleum recovery operations. Enhanced oil recovery (EOR), the process by which 40–80 % of the oil remaining in a reservoir can be recovered after the oil field has stopped producing oil from its own pressure, in becoming increasingly attractive.

Lignosulfonates are over three times cheaper than petroleum sulfonates and are available in larger quantities [15]. Furthermore, lignosulfonates are able to interact synergically with surfactants of petroleum sulfonate, and relatively small quantities of this mixture can establish interfacial tensions as low as 10^{-4} mMm^{-1} between the crude oil and the displacing aqueous solution [164]. The most important observation was that a 60 to 90 % further reduction in the interfacial tension of pure petroleum sulfonate solutions could be achieved by the addition of lignosulfonates [33]. Displacement tests using these surfactant systems could increase tertiary oil recoveries by up to 23 %, as compared to those obtained using pure petroleum sulfonate solutions [33].

Ammonium lignosulfonate has shown promise as a raw material for the production of sacrificial agents. Ozone-oxidized ammonium lignosulfonates, which show increased brine solubility, increased adsorption activity, and high thermal stability, have been successfully used as sacrificial agents, for adsorption protection of the primary surfactant system, and as synergists to increase the effectiveness of the primary surfactant system for EOR [199]. Products are still at the lab-scale development stage, but the field trials for these products will begin in the not-too-distant future.

Lignin has binding properties caused by both chemical and physical intermolecular forces between the lignin molecules and the surface molecules of the particles to be joined. The binding property of lignin is used for soil stabilization and dust control, ore fines briquetting or pelletizing, gypsum and concrete admixture formulations, animal feed pelletizing, etc. Allen and co-workers [6] have reported that the greatest use of lignin sulfonates is for animal feed pelletizing. Lignosulfonates and furfuryl alcohol in the presence of surface activators such as hydrogen peroxide or nitric acid graft on wood and provide effective chemical bridges between the wood surfaces. The composite products made demonstrate strength and water resistance properties comparable to those of conventional phenolic adhesives [179–181].

The special preparation of lignosulfonates or lignin remaining after the steam explosion and enzymatic hydrolysis of wood can be used as a partial substitute for phenol-formaldehyde copolymers in plywood manufacturing. The substitution of lignin for phenol, which may be done for as much as 70%, could help make the phenol-formaldehyde resins, now selling at 1.10 US $ per kg, competitive with less expensive urea formaldehyde adhesives [20].

Lignin seems to be much more promising as an asphalt extender. It can be used as a partial replacement of asphalt in pavement mixtures. Lignin-based extenders for asphalt have good flow and stability characteristics at lignin levels as high as 30% [21,216]. Plywood and particleboard adhesives and asphalt binders are two promising large-volume applications for lignin.

A comparison of the production costs and energy requirements for producing dry lignin from spent pulping liquors to those required for petroleum-derived thermosetting resins, such as phenolic and amino resins, has shown that the use of dry lignins as a full or partial substitute for petro-thermosetting resins is economically feasible [124]. However, the future world market for phenolic and amino resins is relatively small compared to the potential production of technical lignins.

These examples indicate that lignins may be used to synthesize both thermosetting resins and thermoplastics, which have properties very similar to those obtained from petrochemicals. However, the compatibility of lignin with other substances in polymer systems remains to be clarified. It is likely that the high energy required to produce synthetic polymers will favor the future use of lignin as a raw material in the polymer field.

The ability of lignin to form strong covalent bonds with ammonium nitrogen [226] can be used to protect nitrogen in the soil from being washed away or volatilized. Nitrogen would be released when the lignin decayed. Since the process of lignin decay in soil is slow, such 'controlled nitrogen release' could prevent the loss of a considerable amount of nitrogen being used as fertilizer. One process has been patented in which the NH_4HSO_3-based black liquor reacts with ammonium hydroxide in the presence of pressurized air or formaldehyde urea to produce aminated lignin with a 18.5% nitrogen content, which would be useful as fertilizer [3].

Very little has been reported on the medical and biochemical aspects of lignin utilization. Rotstein et al. [191] have reported that feeding lignin, partially and together with lactulose, can completely prevent the formation of cholesterol gallstones, probably because the lignin maintains bile composition within the normal range and increases fecal bile acid excretion. Quinone nitropolycarboxylic acids used as plant-growth stimulators and antichlorosis agents may be prepared from hydrolysis lignin [101]. Compounds with guaiacyl and syringyl structures have been reported to have antimicrobial properties [242]. By the addition of lignin or its derivatives to growing cultures of cellulase-producing fungi, it is possible to produce a desired cellulase preparation that is rich in one particular enzyme and lacks another [221].

7.3 Fragmentation and Chemical Conversion

As shown in Table 8, the fragmentation and chemical conversion of lignin leads to the formation of various phenolic products. Thermal reactions, hydrogenolysis, oxidation, and hydrolysis of lignin have been reviewed by several authors [67,88,224]. The catalytic hydrogenolysis of lignins from aspen, poplar, and sweetgum woods has been studied and the resulting products isolated and identified [206,212,213]. The catalytic hydrogenolysis reactions of lignins show a poor selectivity, and the best yields of monomeric phenols range from 15 to 40%, depending on the type of wood and catalyst used. Loubinoux et al. [141] have suggested the use of nickel boride catalysts to allow for the selective hydrogenolysis of lignins. Catalyst poisoning may occur during hydrogenating of sulfur-containing lignins, and therefore sulfur must be removed prior to hydrogenation.

Hydrolysis of lignin also leads to a mixture of phenolic products, the composition of which depends upon the source of the lignin. The major compound classes in hydrolyzed lignin preparations are phenols, xylenols, catechols, guaiacols, and syringols (Table 9). Vanillin and syringaldehyde are produced respectively by alkaline hydrolysis and by the oxidation of lignin. Alkaline demethylation of lignin leads to the production of dimethyl sulfide (DMS), which is the precursor for dimethyl sulfoxide (DMSO). At present, only vanillin and DMS are produced commercially in rather significant quantities. Vanillin and vanillin derivatives are used mainly as flavoring agents; however, the market is limited. DMS and DSMO are used as solvents, as emulsion matrices for herbicide and insecticide formulations, and also as raw materials in various chemical syntheses [224].

A further process for lignin fragmentation is pyrolysis. Depending upon the feed stock and the pyrolysis temperature, products with various yields are formed. The gaseous products from the pyrolysis of Kraft pine lignin are shown in Fig. 4. The process may be divided into two main stages. Between 120 and 300 C, the phenylpropane side chains are decomposed to formic acid, formaldehyde, H_2O, CO_2, and SO_2. The major decomposition occurs from 300 to 480 C when the principal linkages of lignin are cleaved to phenolics and vapor-phase products [66,152]. Research on lignin pyrolysis focuses mainly upon ways to increase the yields of the ring phenolics that could be of economic importance [6] [p. 49].

The conversion of lignin to phenolic materials has been studied for almost sixty years, but there is still no well-established procedure for harvesting chemically and commercially useful phenolic fractions at a high yield. Although the successful

Table 9. Mono- and dinuclear phenols obtainable through hydrolysis of lignin [88] (numbers refer to formular in Table 10)

Phenols	*Catechols*	*Syringols*
Phenol	2-Hydroxyphenol (Catechol) (83)	2,6-Dimethoxyphenol
2-Methylphenol (*o*-Cresol)	4-Methyl-2-hydroxyphenol (Homocatechol)	4-Methyl-2,6-dimethoxyphenol
3-Methylphenol (*m*-Cresol)	4-Ethyl-2-hydroxyphenol	4-Ethyl-2,6-dimethoxyphenol
4-Methylphenol (*p*-Cresol)	4-Propyl-2-hydroxyphenol	4-Propyl-2,6-dimethoxyphenol
2-Ethylphenol		
3-Ethylphenol	Protocatechuic acid (49)	Syringaldehyde (69)
4-Ethylphenol	Protocatechualdehyde	Syringic acid (57)
4-Propylphenol	Homoprotocatechualdehyde	
3-Methyl-4-ethylphenol		
4-Hydroxyphenol (Hydroquinone)		*Dinuclear phenols*
p-Hydroxybenzoic acid (75)	*Guaiacols*	
p-Hydroxybenzaldehyde		4,4-Dihydroxy-3,3'-dimethoxy-stilbene
	2-Methoxyphenol (Guaiacol) (18)	Dehydrodivanillin
	4-Methyl-2-methoxyphenol	Dehydrodivanillic acid
Xylenols	4-Ethyl-2-methoxyphenol	4,4'-Dihydroxy-3,3'-dimethoxychalcone
	4-Propyl-2-methoxyphenol	4,4'-Dihydroxy-3,3'-dimethoxybenzophenone
2,4-Dimethylphenol (2,4-Xylenol)	(Isoeugenol)	
2,5-Dimethylphenol (2,5-Xylenol)	4-Acetyl-2-methoxyphenol	
2,6-Dimethylphenol (2,6-Xylenol)	(Acetylguaiacone)	
3,4-Dimethylphenol (3,4-Xylenol)	Vanillin (81)	
3,5-Dimethylphenol (3,5-Xylenol)	Vanillic acid (36)	
	Carboxyvanillic acid (60)	
	Carboxyvanillin	

conversion of lignin to phenols has been reported, with conversion yields as high as 40%, the required reaction conditions are often rather severe (e.g., up to 170 atm operation pressure at 450 C). This usually leads to complex mixtures of compounds that are often difficult to recover, which thus limits the economic feasibility of these processes. Hence, future prospects for the production of phenols from lignin is uncertain, at least as long as inexpensive petroleum is available.

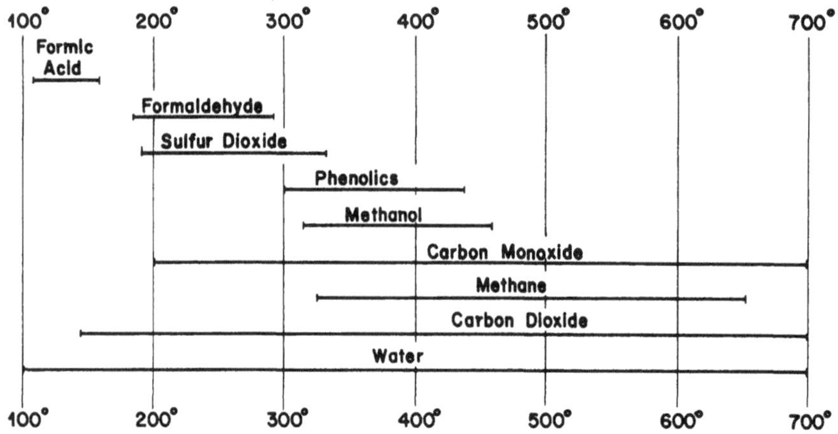

Fig. 4. Gaseous products from the pyrolysis of Kraft pine lignin in a nitrogen atmosphere [66]

However, for the moment at least, it seems that a two-stage process consisting of hydrocraking, followed by a hydrodealkylation, is most promising for the conversion of lignins to phenolic compounds [200]. A 'lignol process' was developed on this basis and was estimated capable of converting Kraft lignin into about 20% phenol, 14% benzene, 13% fuel oil, and 29% fuel gas [31].

A capital investment of about 45 million dollars (1980) would be required for a plant processing 485 tons per day of Kraft lignin, with a return investment estimated at 25%. However, whether there might be any interest from the paper companies in using this process is not clear at this time.

8 Biodegradation and Transformation

The tremendous amount of lignin produced annually from photosynthesis on our planet is balanced by the amount of lignin that is decomposed by microorganisms. Although this phenomenon plays a central role in the earth's carbon cycle, our current understanding of it is fairly poor and is yet susceptible to scientific explanation. Actual research on lignin degradation did not start before the chemistry of lignin attained the stage that could allow such studies. Due to recognition of the value of lignin as a possible alternative and renewable source of chemical feedstocks, research on lignin biodegradation has intensified during the last few years. However, in spite of the strong efforts now being made in this area, because of the complexity of the subject, there are still many unanswered questions that will keep scientists busy for at least the next decade. Comprehensive books on lignin have been published by Kirk et al. [121] and by Crawford [43]. The following section should complement and update these publications.

8.1 Substrates for the Study of Lignin Degradation

The most challenging task in the study of lignin degradation is the choice of material and methods. The composition and structure of lignin is so different and susceptible to further changes during the performance of experiments that it is almost impossible to have a standard lignin preparation for use in biological degradation studies. However, careful selection of lignin types and experimental processes can help scientists obtain more reliable results, although with recognizable or possibly predictable deviations. Two categories of lignin are usually used for the study of degradation: extractive lignins and synthetic compounds with certain structural similarities to lignin.

8.1.1 Extractive Lignins

Details on extraction procedures, property and degradation of these preparations are beyond the scope of this review (see [8,43]). In this section, the main characteristics of the extractive lignins used most often are described from the basis of substrate material for microbial growth and lignin transformation studies.

Acidolysis lignin or dioxane lignin: Lignin is extracted from plant tissues by using a mild acid hydrolysis (0.2M HCl in aqueous dioxane) at ambient temperatures [174]. This lignin is reported to have only few carbohydrate impurities and is a convenient substrate for the isolation and study of lignin degrading microorganisms [174]. However, acidolysis procedures are known to affect numerous lignin substructures (see Sect. 6.4).

Brauns' native lignin: This lignin is prepared by the extraction of lignocellulosic material using 95% solution of ethanol in water. The obtained yield is very low because the bulk of the lignin is insoluble in ethanol and only a small portion is soluble in ethanol-water, for example, about 1% of spruce lignin. The solubilized lignin probably contains only very low molecular fractions and is generally considered not representative of the bulk of lignin in the extracted tissues.

Cellulase lignin: Cellulase lignin is prepared from milled wood after the removal of polysaccharides using a commercially available cellulase-hemicellulase mixture. The enzyme mixture usually cannot remove all of the carbohydrates so the lignin preparation may have 12–14% carbohydrate contamination. The yield is large. Of the different preparations thus far available, this lignin is most representative of the bulk of lignin. Therefore, the results obtained when using it can be better interpreted in in-situ situations.

Enzymatically liberated lignin: In this process, woody tissue is inoculated with brown rot fungi. The fungi decompose the carbohydrate part of the wood, and the residue is the so-called 'brown-rotted' lignin. This lignin, compared to sound lignin, however, is altered, having less methoxyl and more α-carbonyl content. Enzymatically liberated lignin also undergoes other fungus-caused chemical modificatons that are not yet well understood.

Klason lignin: Klason lignin refers to the material left after cold, 72% sulfuric acid treatment of lignocellulosic material. The extreme conditions under which this lignin is prepared causes the material to be highly condensed and resinous. This treatment is still used as a quantitative method for measuring the acid-insoluble lignin content of lignocellulose (see Sect. 8.2.4). Nevertheless, it should not be used for biodegradation studies because intensive alteration and condensation affects the lignin during this treatment.

Kraft lignin and lignosulfonate: These lignins are readily available and therefore are widely used as experimental lignins for studies of lignin biodegradation. However, they are not pure. They contain non-lignin derived materials and have low molecular weight degradation products of lignin. Purification and fractionation steps have to be undertaken to obtain a rather homogeneous preparation. It has been suggested that an organic-soluble, ether-insoluble fraction of Kraft lignin (Indulin) can be used as a standard lignin preparation in research and for analytical work [148]. This fraction of Kraft lignin has been shown to contain molecules with molecular weights above 3000.

Lignin-carbohydrate complex (LCC): Lignocarbohydrate complexes have been used for biodegradation studies by only a few groups. Milstein et al. [156] tested different fungi for their ability to degrade LCC and for their growth on aqueous hydrolyzate liquors of wheat straw. Extractions were made by percolation or by autoclaving. It should be kept in mind that encrustation of lignin in the lignin-carbohydrate complex can affect lignin biodegradability, especially in the cases of non-polysaccharide-degrading microorganisms [174] and rumen hemicellulases [26].

Milled wood lignin (MWL): Milled wood lignin is usually prepared according to the procedure of Björkman [22]. The wood is ball-milled in a non-lignin-swelling solvent to remove extraneous components. Lignin is then extracted using dioxane-water, sometimes followed by a multi-step purification process. Only minor changes are supposed to occur in the structure of lignin during this procedure. Therefore, MWL is one of the best lignin preparations for microbial studies. However, the yield is low. At the most, 50% of the lignin of the balled-milled wood can be extracted. This low yield may be due to lignin-carbohydrate bonds that can predominate and determine the solubility of the lignin [183]. Vigorous grinding can destroy a part of the covalent bonds in a lignin-carbohydrate complex, eliminate the hinderance to lignin solubilization, and increase the yield of extracted lignin. Grinding, however, cannot eliminate all of the LCC linkages. It may even cause structural damage to the lignin itself. Björkman lignin typically contains a small percent of carbohydrates, the complete removal of which has not yet been achieved. The amount of the resulting carbohydrate impurities apparently depends upon the extent of milling and the conditions used for extraction, as well as upon the plant species. Björkman lignin from spruce, for example, is reported to have a much lower carbohydrate content than corresponding preparations from birch [149].

Milled wood lignin is not considered to be representative of whole lignin in wood. It originates primarily from the secondary wall of the plant cell tissue [230]. Further milling affects the wood ultrastructure and indirectly influences lignin biodegradability. Higher degradation rates may be achieved by increasing the extent of the milling. As a consequence, the biodegradaton of MWL does not represent in-situ lignin degradation.

Labeled plant tissues (see Sect. 8.2.5) are also used for the preparation of MWL. For this purpose, the lignifying plants are fed by the lignin precursors such as L-phenylalanine. One commercially available precursor is L-[U-^{14}C] phenylalanine. The problem accompanying the application of L-phenylalanine is that it incorporates not only into the plant lignin, but also into the plant protein. Since the plant protein cannot be completely removed using available chemical methods, the production of a protein-free lignin is almost impossible (see Sect. 6.3). The protein is also usually easier to be degraded than the lignin and it takes part in the biodegradation assays, which could mislead investigators to select microbes that are, in fact, not lignin degraders. It has been reported that deaminated lignin precursors such as *trans*-cinnamic acid, *p*-coumaric acid, and ferulic acid cannot be incorporated into plant protein and should be used in place of L-phenylalanine. These compounds, however, are rather expensive. Pometto and Crawford [184)] have suggested a method for the production of *trans*-[U-^{14}C] cinnamic acid from L-[U-^{14}C] phenylalanine, using a commercially available preparation of phenylalanine ammonia-lysase.

8.1.2 Lignin Model Compounds

The complexity of lignin polymers, the disadvantages of extractive lignins, and the problems concerning the qualitative and quantitative determination of lignins have compelled scientists to use low molecular compounds related to structural elements in lignin polymers for lignin biodegradation studies. The successful application of model compounds in other aspects of lignin chemistry is another reason for choosing this approach. Because of the frequent occurrence of β-aryl ether linkages in the structure of lignin (Table 5), the preferred lignin model compounds are those having such linkages in their structures. However, the relevance of much information obtained on degradation reactions using lignin model compounds to the actual degradation of the lignin polymer is yet to be answered. In the following sections, model compounds that have been used most often in recent years are described.

Dehydrogenative polymerizate (DHP): This polymer is synthesized in vitro by oxidative polymerization of immediate precursors of lignin, *p*-hydroxycinnamyl alcohols, in a peroxidase H_2O_2-catalized reaction. DHP is formed after a continuous series of phenol-coupling reactions between peroxidase-generated cinnamyl alcohol radicals.

Enzymic dehydrogenation of *p*-coumaryl alcohol has been described by Nakatsubo [160)] and compared with that of coniferyl alcohol. DHPs are very good lignin model polymers and their chemical and physical properties have been well characterized. The lignin precursors may also be labeled in different positions for radioisotopic studies. However, [^{14}C]DHPs are somewhat difficult and expensive to prepare. For lignin biodegradation research, Crawford et al. [45)] have recommended a ^{14}C-labeled polymer of *o*-methoxyphenol (polyguaiacol) as a useful model polymer, particularly regarding the biphenyl structures of lignin. This compounds is easier and less expensive to prepare than DHPs. The polymer has an average molecular weight of between 5000 and 15000 and is primarily composed of o-o and p-p-linked guaiacol moieties.

Table 10 Major di-and monomeric lignin model compounds used for the elucidation of lignin degradation pathways by microorganisms

Lignin model	Name	Organism	Ref.
1 *Dilignols*			
1.1 *Arylglycerol*-β-aryl ether (β-o-4) type structure			
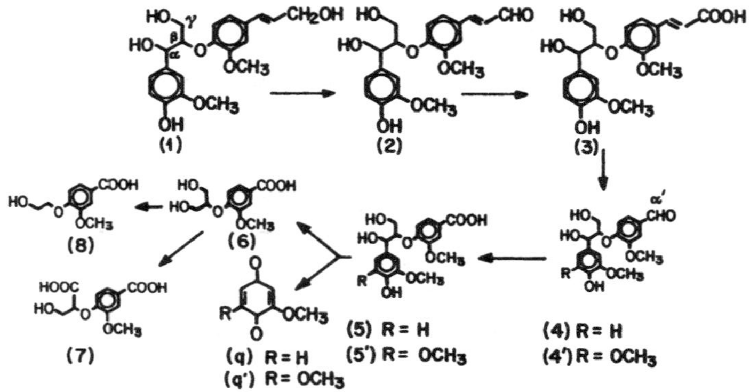	(1) guaiacylglycerol β-coniferyl ether, (2) guaiacylglycerol-β-coniferaldehyde, (3) guaiacylglycerol β-ferulic acid ether, (4) guaiacylglycerol β-vanillin ether, (4′) syringylglycerol β-vanillin ether, (5) guaiacylglycerol -β-vanillic acid ether, (5′) syringylglycerol β-vanillic acid ether, (6) glycerol β-vanillic acid ether, (7) glyceric acid 2-vanillic acid ether, (8) ethyleneglycol-mono vanillic acid ether, (9) 2-methoxy-*p*-benzoquinone, (9′) 2,6-dimethoxy-*p*-benzoquinone	*Fusarium solani*	97, 104, 115, 116)
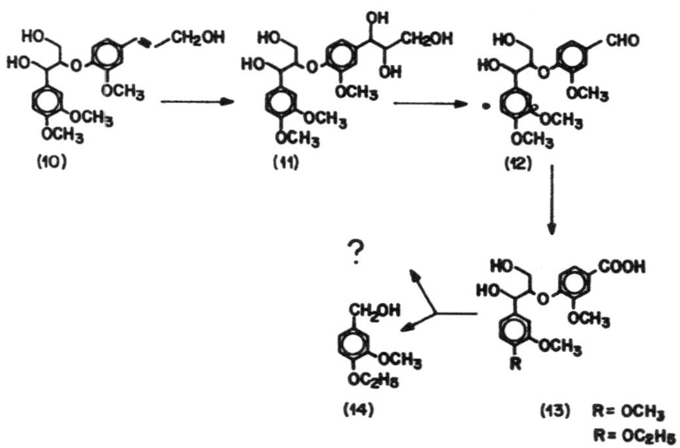	(10) veratrylglycerol β-coniferyl ether, (11) veratrylglycerol β-guaiacylglycerol ether, (12) veratrylglycerol β-vanillin ether, (13) veratrylglycerol β-vanillic acid ether, (14) 4-ethoxy-3-methoxybenzyl alcohol	*Phanerochaete chrysosporium*	96, 98)

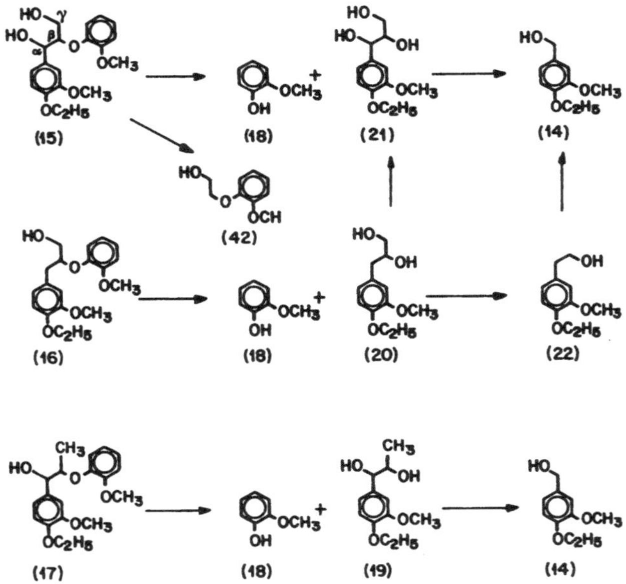	(15) 4-ethoxy-3-methoxyphenyl-glycerol β-guaiacyl ether, (16) 1-(4′-ethoxy-3′-methoxyphenyl)-2-(2″-methoxyphenoxy)-3-hydroxypropane, (17)1-(4′-ethoxy-3′-methoxyphenyl)-2-(2″-methoxyphenoxy)-1-hydroxypropane, (18) guaiacol, (19) 1-(4′-ethoxy-3′-methoxyphenyl)-1,2 dihydroxy-propane, (20) 1-(4′-ethoxy-3′-methoxy-phenyl)-2,3 dihydroxypropane, (21) 4-ethoxy-3-methoxyphenyl-glycerol, (22) 1-(4′-ethoxy-3′-methoxyphenyl)-2-hydroxy-ethane	*P. chrysosporium* [57]
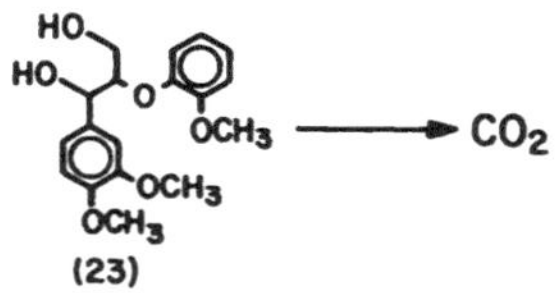	(23) veratrylglycerol-β-guaiacyl ether	*P. chrysosporium* [228]
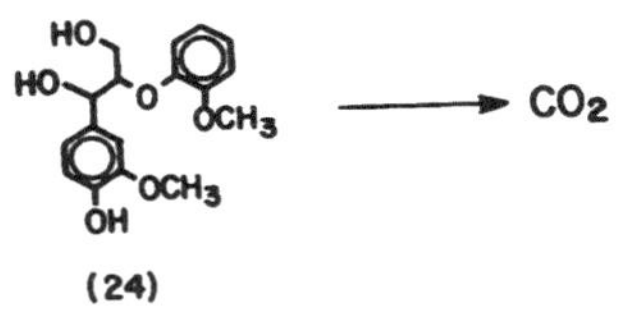	(24) guaiacylglycerol-β-guaiacyl ether	*P. chrysosproium* [228]

Table 10 (continued)

Lignin model	Name	Organism	Ref.
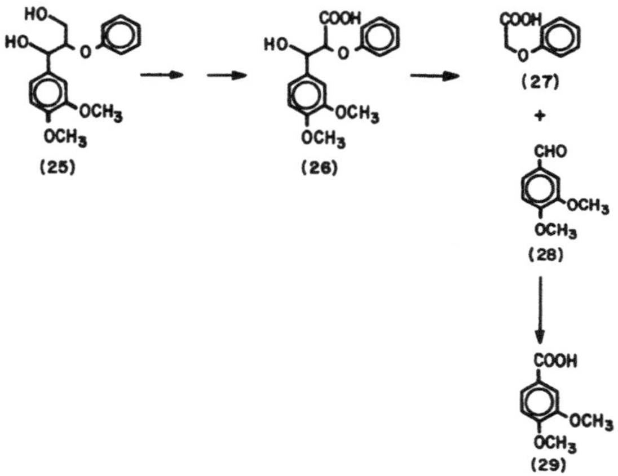	(25) veratrylglycerol-β-phenyl ether, (26) 2-phenoxy-3-hydroxy-3-(3′, 4′-dimethoxy phenyl)-propionic acid, (27) phenoxy-acetic acid, (28) 3,4-dimethoxybenzaldehyde, (29) 3,4-dimethoxybenzoic acid or veratric acid	*Arthrobacter*, *Brevibacterium*, *Corynebacterium*, *Nocardia*, *Rhodococcus*	[186]
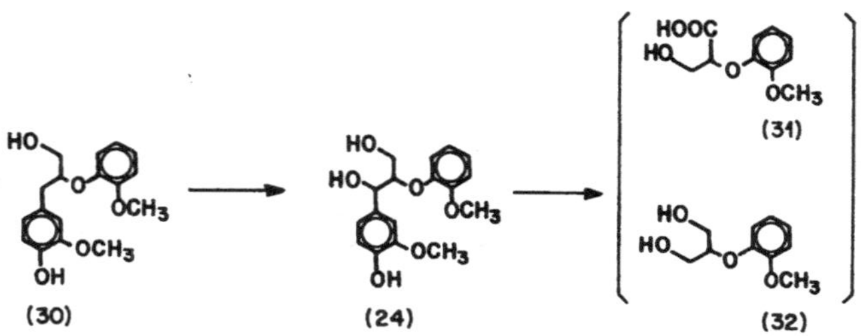	(30) α-deoxyguaiacyl-glycerol-β-guaiacyl ether, (31) 3-hydroxy-2-(2′-methoxyphenoxy) propionic acid, (32) 2-(2′-methoxyphenoxy)-1,3 propanediol	*P. chrysosporium*	[76]

1.2 *Arylglycol-β-aryl ether type structure*

Scheme	Compounds	Organism	Ref.
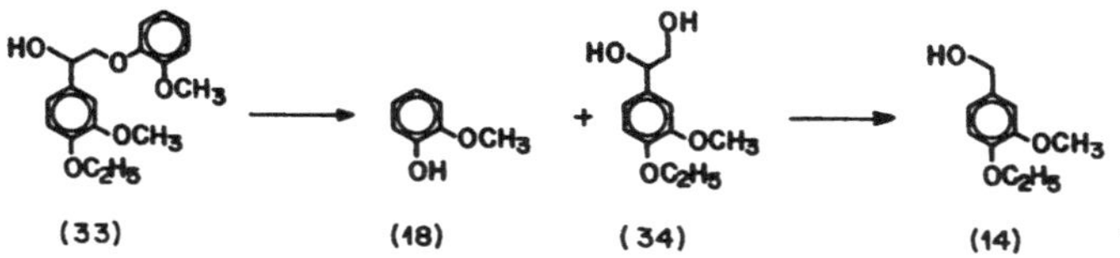	(33) 4-ethoxy-3-methoxyphenylglycol-β-guaiacyl ether, (34) 1-(4′-ethoxy-3′-methoxy-phenyl)-1,2-dihydroxy ethane	*P. chrysosporium*	58)
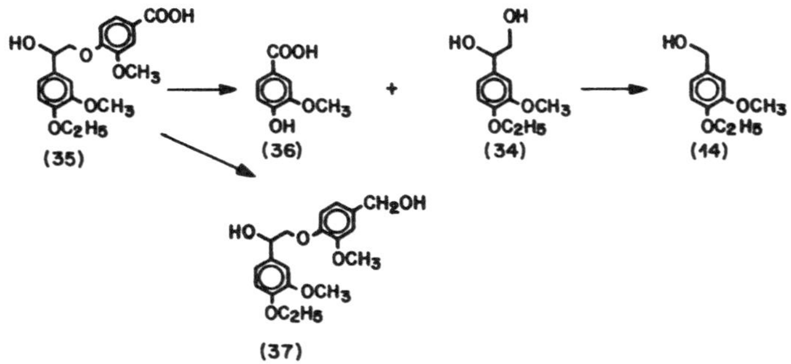	(35) 4-ethoxy-3-methoxyphenylglycol-β-vanillic acid, (36) vanillic acid, (37) 4-ethoxy-3-methoxyphenylglycol-β-vanillyl alcohol ether	*P. chrysosporium*	58)
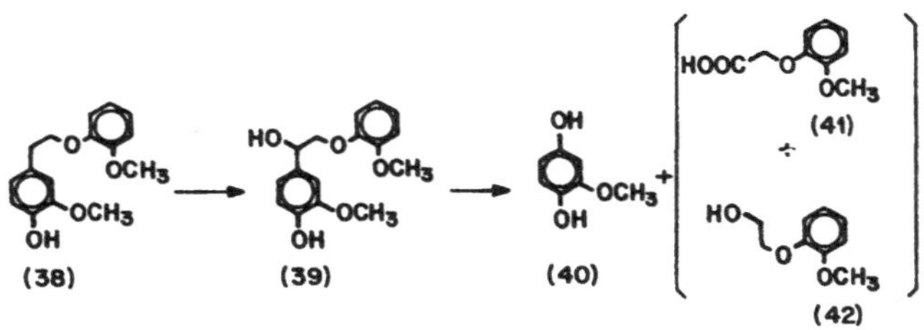	(38) α-deoxyguaiacylglycol-β-guaiacyl ether, (39) guaiacylglycol-β-guaiacyl ether, (40) methoxyhydroquinone, (41) 2-(2′-methoxy-phenoxy) acetic acid, (42) 2-(2′-methoxy-phenoxy) ethanol	*P. chrysosporium*	76)

Table 10 (continued)

Lignin model	Name	Organism	Ref.
1.3 *Phenylcoumaran type structure* 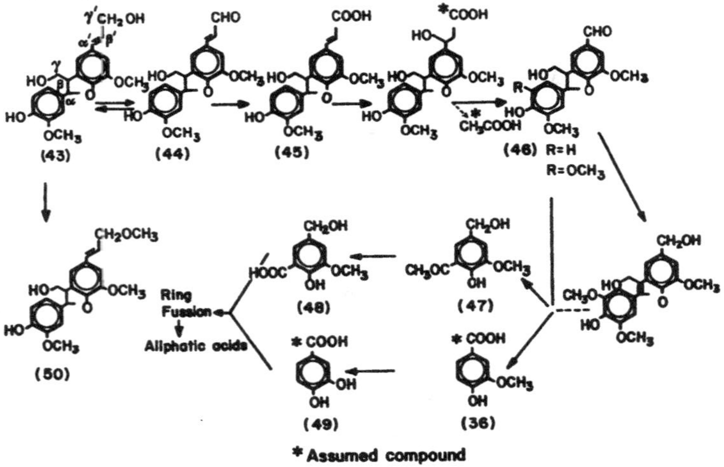	(43) dehydrodiconiferyl alcohol, (44) 2-(4-hydroxy-3-methoxyphenyl)-3-hydroxymethyl-5-(2-formylvinyl)-7-methoxycoumaran, a phenylcoumaran-γ′-aldehyd, (45) 2-(4-hydroxy-3-methoxyphenyl)-3 hydroxymethyl-5-(2-carboxyvinyl)-7-methoxycoumaran, a phenylcoumaran-γ′-carboxylic acid, (46) 2-(3-hydroxy-2-methoxyphenyl)-3-hydroxymethyl-5-formyl-7-methoxycoumaran, a phenylcoumaran-α′-aldehyd, (47) 5-acetylvanillyl alcohol, (48) 5-carboxyvanillyl alcohol, (49) protocatechuic acid, (50) 2-(4-hydroxy-3-methoxyphenyl)-3-hydroxymethyl-5-(3-methoxyallyl)-7-methoxycoumaran, a γ′-methyl ether of dehydrodiconiferyl alcohol	*F. solani*	114)

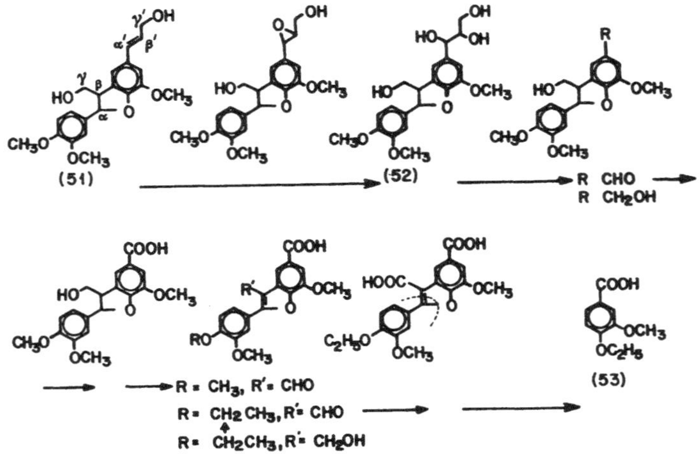	(51) 4-*o*-methyl dehydrodiconiferyl alcohol, (52) 4-*o*-methyl dehydrodiconiferyl glycerol, (53) vanillic acid ethyl ether	*P. chrysosporium*	162, 210)
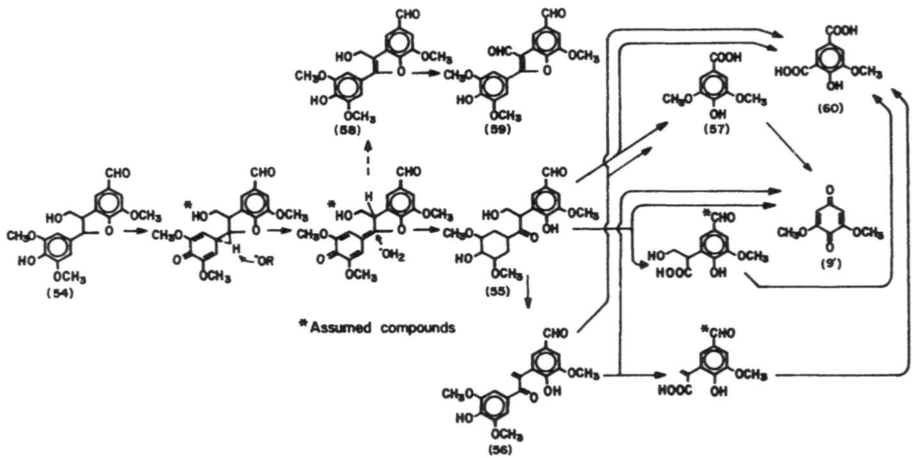	(54) 5-formyl-3-hydroxy methyl-2-(4-hydroxy-3,5-dimethoxy-phenyl)-7-methoxycoumaran, a phenylcoumaran α′-aldehyd, (55) 2-(5-formyl-2-hydroxy-3-methoxy-phenyl)-3-hydroxypropiosyringone, (56) 2-(5-formyl-2-hydroxy-3-methoxyphenyl)-2-methylenaceto-syringone, (57) syringic acid, (58) 5-formyl-3-hydroxymethyl-2-(4-hydroxy-3,5-dimethoxyphenyl)-7-methoxycoumarone, (59) 3,5-diformyl-2-(4-hydroxy-3,5-dimethoxy-phenyl)-7-methoxycoumarone, (60) 5-carboxyvanillic acid	*P. chryso-sporium*	220)

Table 10 (continued)

Lignin model	Name	Organism	Ref.
1.4 *Pinoresinol (β-β′) type structure* 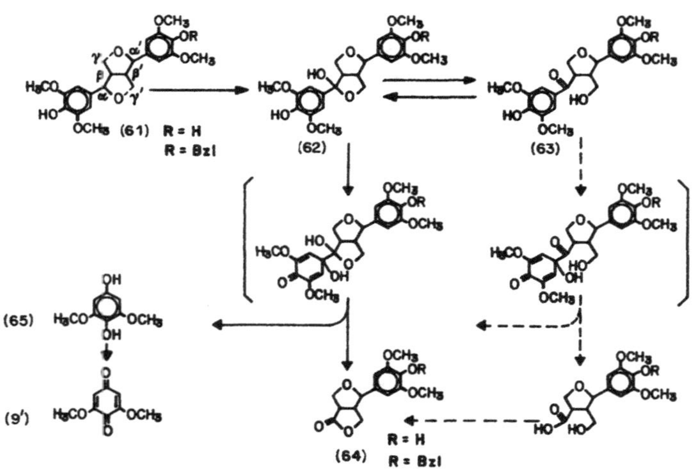	(61) syringaresinol, (62) α-hydroxysyringaresinol, (63) 3-hydroxymethyl-2-(4-benzyloxy-3,5-dimethoxyphenyl)-4-(4-hydroxy-3,5-dimethoxybenzoyl)-tetrahydrofuran, (64) 6-oxo-2-(4-benzyloxy-3,5-dimethoxyphenyl)-3,7 dioxabicyclo-[3,3,0]-octane, (65) syringalcohol	*F. solani*	99,102,103,112)

1.5 *Diarylpropane-1,3-dial (β-l) type structure*

Structure	Compounds	Organism	Ref.
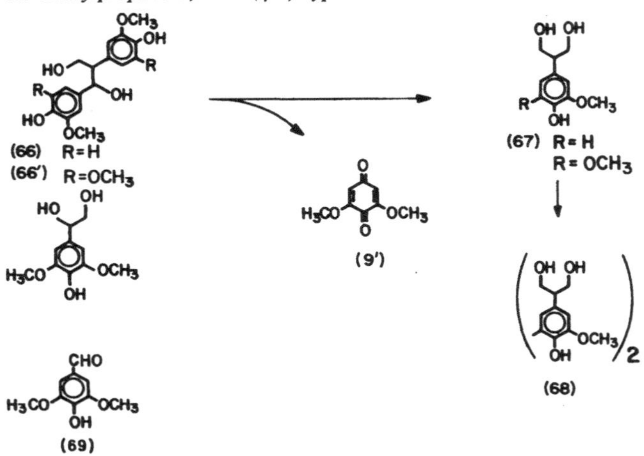	(66) 1,2-diguaiacylpropane-1,3-diol (66′) 1,2-disyringylpropane-1,3-diol, (67) 2-guaiacylpropane-1,3-diol, (68) biphenyl dimer of 67, (69) syringaldehyde	*F. solani*	163)

2 *Monolignols*

2.1 *Phenyl-propanoid type structure*

Structure	Compounds	Organism	Ref.
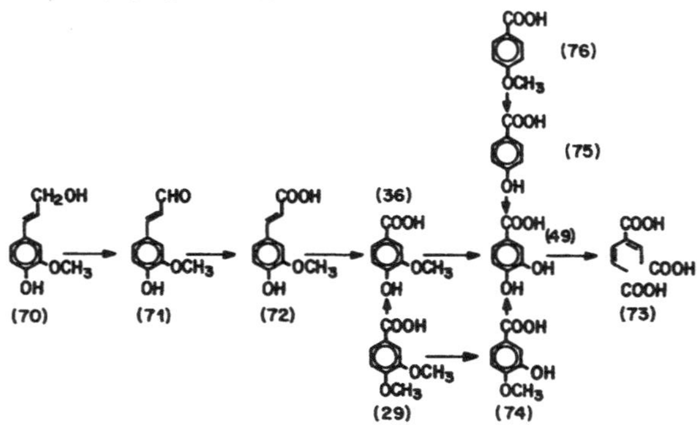	(70) coniferyl alcohol, (71) coniferyl aldehyde, (72) ferulic acid, (73) β-carboxy-*cis*, *cis*-muconic acid, (74) isovanillic acid, (75) 4-hydroxybenzoic acid, (76) 4-methoxybenzoic acid	*Nocardia* sp.	54)

Table 10 (continued)

Lignin model	Name	Organism	Ref.
2.2 *Phenylalanine*			
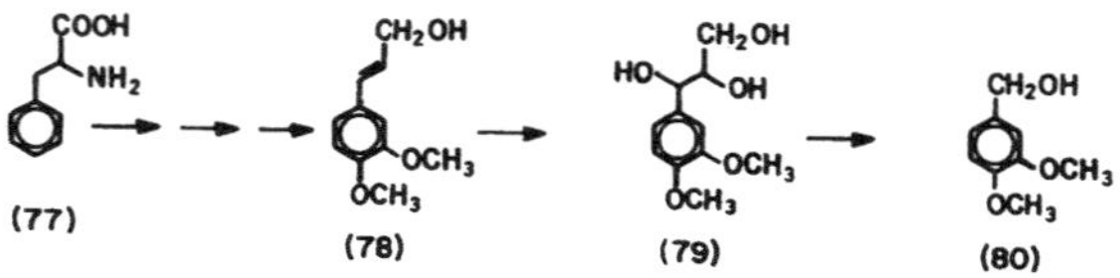	(77) phenylalanine, (78) 3,4-dimethoxycinnamyl alcohol, (79) veratryl glycerol, (80) veratryl alcohol,	*P. chrysosporium*	[210]
2.3 *Vanillic acid*			
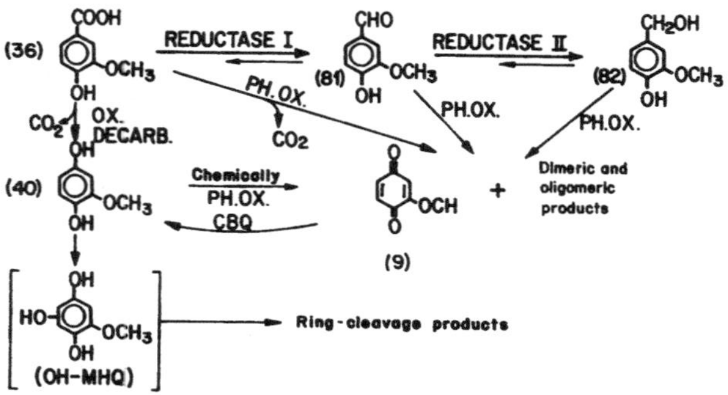	(81) vanillin, (82) vanillyl alcohol,	*P. chrysosporium*	[9]
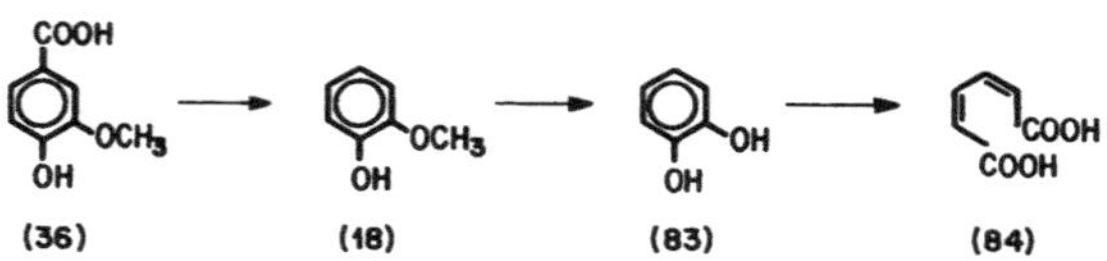	(83) catechol, (84) *cis*, *cis*-muconic acid	*Streptomyces setonii*	[185]

Oligolignols: The general methods for the synthesis of dimeric lignin model compounds such as guaiacylglycerol-β-guaiacyl and guaiacylglycerol-β-coniferyl ethers, 1.2-diarylpropane-1,3-diols, phenylcoumarans, as well as that of some triligols, have also been described [160]. A modified procedure was described by Lawrence et al. [134] for a small-scale and efficient synthesis of 3-methoxy-4-hydroxy-α-(2-methoxyphenoxy)-β-hydroxypropiophenone from commercially available [Ring-U-^{14}C]-phenol with an overall yield of about 9% in a ten-step sequence. Synthesis of lignin model compounds of the 1.2-diaryl-1,3-propanediol type via benzyloxy rearrangement of chalcone was reported by Kristersson and Lundquist [125]. Dehydrodivanillin contains the important 5,5'-biphenyl linkage (Table 5) and may be used as another dimeric lignin model. However, Crawford et al. [44] using several mutants of *Streptomyces viridosporus* have shown that dehydrodivanillin is not a relevant model compound for the study of lignin degradation. The mutant strain, as compared to the wild type strain, lost its ability to catabolize dehydrodivanillin, while it retained an undiminished ability to degrade Douglas-fir lignin.

Monomers: Monomers have also been used in connection with lignin degradation, especially for bacteria. Vanillic (4-hydroxy, 3-methoxybenzoic) and ferulic (4-hydroxy, 3-methoxycinnamic) acids, for example, are the key intermediates in the degradation of lignin and lignin-related substrates by microorganisms. Therefore, these acids have been used for the elucidation of enzyme mechanisms involved in lignin degradation [9,79]. Lignin monomers such as coniferyl aldehyde, *p*-hydroxycinnamaldehyde, and sinap aldehyde have been sythesized [129].

The major lignin model substrates used for the study of lignin degradation pathways by different microorganisms are listed in Table 10. It is worth mentioning that the study of the biodegradation of lignin model compounds does not shed light on the initial steps involved in the breakdown of lignin polymers.

8.2 Methods for the Study of Lignin Structure and Degradation

The most important restriction for the progress in the study of lignin structure in the last 50 years has been inadequate methodology. The complexity and divergency of the lignin molecule and the fact that it cannot be detected using a single chemical method have slowed progress in lignin research. Several methods have been suggested for the determination of lignin in plant materials, in soil and natural waters, and in materials from the paper industry. Some of these methods have been adopted to study microbial lignin decay. Lignin can be determined either directly or indirectly. In direct methods, such as those based on mineral acid or alkali extractions, lignin is isolated directly and determined either as an insoluble residue or in a solution. In indirect methods, researchers assayed for a characteristic group or a characteristic chemical reaction of lignin.

The most frequent structural changes occurring in the lignin macromolecule during or after microbial attack is the oxidative splitting of side chains between α and β carbons, which produces aromatic acid residues; the splitting of β-arylether linkages followed by oxidative modification of side chains; the degradation of aromatic nuclei to carboxylic acid groups; and many other yet undiscovered changes and modifications (see Sect. 8.6). Detection of all these events is exceedingly difficult and is certainly beyond the capability of many chemical and biochemical laboratories. Several approaches and methods have to be undertaken at the same time in order to allow a correct and reliable understanding of lignin reactions.

8.2.1 Chemical Degradation and Fractionation

Many degradation reactions of lignin have been employed for the characterization of decayed lignin. Oxidation-based lignin determination techniques, such as 'chlorine'- and 'Kappa'-number standard techniques, have found world-wide acceptance in the pulp and paper industries, but the applicability of these methods to biological systems has to be tested [223]. Oxidative degradation of methylated, ethylated, and acetylated lignin leads to the formation of different ketons and aromatic acids such as veratric acid, isohemipinic acid, and dehydrodiveratric acid. The yield of these degradation products can provide certain evidence of the structural changes occurring to lignin molecules during decay. Ketones are usually derived from β-O-4 structures. Hence, lower yields of ketones from decayed lignin, as compared to sound wood lignin, would mean that such linkages have been destroyed during decay. The decayed lignin is reported to produce less veratric acid and more isohemipinic acid after chemical oxidative degradation [30].

Successive extraction and fractionation of wood by different solvents, such as ligroin, chloroform, acetone, methanol, aqueous dioxane, etc., and comparisons of the yield of each extracted fraction to that of undecayed wood also provides valuable information. The fractionation process can be extended by separating the extracted material into acidic and phenolic parts so that each part could then undergo further purification.

8.2.2 Elemental Composition and Functional Group Analysis

Lignin is mainly composed of C, H, and O elements. An analysis of these elements, together with a determination of certain functional groups of lignin, provides useful information about lignin composition (Table 11). The methoxyl content has been recognized as an essential criterion for characterizing lignins. Other lignin functional groups, such as conjugated carbonyl, carboxylic, phenolic, and aliphatic hydroxyl groups, have also been determined (Table 12). However, the methods based on physicochemical properties of lignin, such as assays based on methoxyl determination, cannot differentiate between degradation to CO_2 and simple demethylation.

Table 11. Elemental and methoxyl analyses and C_9-unit formulae for sound (not decayed) and white-rotted spruce lignins [120]

Lignin sample	C	H	O	OCH_3	C_9-unit formulae	C_9-unit Mol. wt
Sound[a]	62.85	6.08	31.07	15.11	$H_{8.66}O_{2.75}(OMe)_{0.92}$	189.2
Fungus-degraded[b]	57.97	4.70	37.23	11.33	$H_{7.26}O_{3.95}(OMe)_{0.74}$	199.4

[a] Milled wood lignin;

[b] Purified from wood which had been decayed to 50% weight loss by *Coriolus versicolor*

Table 12. Property changes in lignins from different woods caused by several white-rot fungi [120]

Property	Change[a] Increase	Decrease	Method of analysis[b]
Carboxyl content	+		C, UV, IR, PMR
Hydroxyl content			
phenolic	+[c]	+	C, UV, PMR
aliphatic		+	C, PMR
Carbonyl content	+		UV, IR, PMR
Hydrogen/carbon		+	C
Oxygen/carbon	+		C
Methoxyl/carbon		+	C
Yield of methoxylated aromatic acids on oxidative degradation after methylation[d]		+	C
Yield of principal acidolysis products[e]		+	C
Yield of major products on nitrobenzene oxidation[f]		+	C

[a] Degraded lignins were compared to sound lignins
[b] C = various chemical methods: UV, IR, PMR = ultraviolet, infrared, and proton magnetic resonance spectroscopy, respectively
[c] Hata (1966) found an increase in phenolic hydroxyl content and obtained variable results with aliphatic hydroxyl content
[d] The major product is 3,4-dimethoxybenzoic (veratric) acid from gymnosperm lignin. From angiosperms, veratric acid and tri-O-methyl gallic acid are dominant
[e] The major product from sound lignin is 3-hydroxy-1-(4-hydroxy-3-methoxyphenyl)-2-propanone. Vanillic acid is the major product from white-rotted lignin
[f] Vanillin (gymnosperms) and vanillin + syringaldehyde (angiosperms) are the major products

The presence of large amounts of conjugated carboxylic acid other than aromatic acids, a deficiency in methoxyl content, an increase of oxygen, and a decrease of the hydrogen content in lignin have been interpreted as an indication of an oxidative ring cleavage [120]. White-rotted lignin exhibits a distinct increase in O/C content and an increase in carboxyl and carbonyl groups with a corresponding decrease in H/C content (Table 12). With respect to hydroxyl content, the reported data are discrepant. Hall et al. [86] have indicated that there is very little change in the hydroxyl content.

8.2.3 Spectroscopy

Spectroscopical methods have shown to be good for the determination of lignin and lignin derivatives. The characteristic absorption spectra of lignin has been used for quantitative and qualitative determinations of lignin [5,105,133] and lignosulfonate [154,211], as well as for the study of delignification of wood fiber cell walls during pulping [70].

In UV spectrophotometry, the typical absorbance maximum of lignins of about

280 nm should be attributed to the π-π^* transition in the aromatic rings of the polymer. The decrease in the absorptivity coefficient at this wavelength can be due to the cleavage of the aromatic ring structures within the polymer. This method suffers from the disadvantage of interference from non-lignin compounds with considerable absorbance values at UV range. Samples have to be extracted and purified. The effects of culture medium composition, pH, biomass, metabolites, and several lignin-related aromatic compounds on the UV absorption of lignin solutions have been reported [105]. For the extraction and purification of lignin dioxane-water was suggested [105,108].

Besides UV, IR spectroscopy has also helped to estimate certain functional groups such as conjugated carbonyl, conjugated and total phenolic hydroxyl groups, as well as aromatic nuclei content, carboxylic acids and esters, etc. IR spectra of Björkman lignin, humic compounds [208], Kraft lignin [148], lignin isolated from spent pulping liquors [150], and di- and trilignols [160] are now available. UV and IR spectra of the lignincarbohydrate complex have been studied [13]. IR spectroscopy has also been used to study the bonds between nitrogen and lignin [226] and to study aminolignins obtained by treatment of hydrolysis lignins with ammonia [50]. IR spectroscopy may be applied for demonstrating differences between softwood and hardwood lignins [170].

Lignin and lignin model compounds have fluorescence properties. Fluorescence-spectroscopy has been used for the analysis of lignin, ligninsulfonate, and related compounds [147,154]. However, the presence of stilbene structures and arylconjugated carbonyl groups strongly influences the fluoresence properties of lignins. It has therefore been suggested that a borohydride reduction should be performed prior to the quantitative analyses of lignins to give more reliable results [147]. A method has been developed for coupling high molecular weight lignosulfonate to a fluorescent dye to provide conditions so that degradation products do not enter the cell and hence remain in the culture to be fluorimetrically detected [82]. The main advantage of this method is that it allows for the monitoring of the first events in lignin catabolism and therefore may be a basis for the study of in vitro lignin degradation.

Characterization of lignin preparations by ^{13}C NMR and ^{1}H NMR spectroscopy is one of the most effective techniques for their examination. Schäfer et al. [203] have reported the use of cross-polarization and magic-angle spinning (CPMAS) ^{13}C NMR to assay chemical changes brought about by fungal transformation. Chua et al. [35] used ^{13}C NMR to characterize different fractions of milled wood lignin before and after fungal attack. This technique has also been used to investigate bacterial and fungal bioalteration of ^{13}C-enriched DHP of coniferyl alcohol [55]. The structure of milled wood lignin from birch and spruce, industrial lignins, and the analysis of carbohydrates in lignin preparations were investigated by ^{1}H NMR spectroscopy with a 270 MHz instrument [144,146,149,150]. The high resolution of the 270 MHz indicated that ^{1}H NMR spectroscopy with a 270 MHz instrument could be used to a good advantage for comparing lignins of different origins and for studying the chemical changes of lignins caused by various treatments [144]. NMR spectra of different lignin model compounds have been reported by Nakatsubo [160]. ^{1}H NMR and ^{13}C NMR spectroscopies have also helped demonstrate the incorporation of chlorinated anilines, the primary metabolites of a number of widely used pesticides and other chemicals, into lignin [219].

8.2.4 Gravimetry

Gravimetric methods based on acid- insolubility of lignin are neither fully accurate nor specific and can be used as criteria for lignin decomposition only when lignin decomposition or depletion from samples is great enough to leave no reasonable doubt. The condition for acid hydrolysis in the Klason lignin determination method is sufficient to hydrolyze polysaccharrides, but proteins are only partially dissolved. Whitmore [231)] has found 9% protein in the Klason lignin preparation from the cell walls of *Pinus elliottii*. Furthermore, use of the Klason method does not work well with hardwood lignin, which contains significant fractions of acid-soluble lignin. Gravimetric and spectrophotometric methods of lignin determination have been used to compare for a variety of roughage sources in connection with their digestibility by ruminants [159)]. It was found that the choice of method could affect the apparent lignin digestibility of forage for ruminants.

8.2.5 Radioisotopy

Radioisotopic methods have been applied very successfully in the study of lignin biodegradation (see also 8.1). This method, however, measures only the final product of total lignin degradation, i.e., CO_2. Determination of $^{14}CO_2$ generally provides an underestimation of lignin degradation since it does not account for the incorporation of ^{14}C into the cell mass and into other lignin-derived compounds. Furthermore, some microorganisms may degrade lignin without converting it substantially to CO_2, as occurs in humification. Crawford and Crawford [41)] have suggested that the sum of $[^{14}C]$ lignin converted to $^{14}CO_2$ and water-soluble ^{14}C is more indicative of ligninolytic activity. In general, if an organism can make greater than 3% decomposition of $[^{14}C]$-DHPs or greater than 10% degradation of ^{14}C-plant lignins (as percentage converted to $^{14}CO_2$ or $^{14}CO_2$ plus water-soluble ^{14}C), the organism should be assumed to be a significant modifier of lignin [241)]. Radioisotopy has also been used for H-labeled synthetic lignins to study the stereochemical changes occurring to lignin during biological degradation [168)].

8.2.6 Microscopy

Lignin in the wood fiber cell wall has been studied by different microscopical methods such as UV microscopy [25)], transmission electron microscopy (TEM) [192)], and scanning electron microscopy (SEM) [61)]. Light microscopy and planimetry have been used to determine the lignified area in cell tissue [218)]. Yang and Goring used ultraviolet microscopy to measure the phenolic hydroxyl content of lignin in situ in wood [237)]. However, the obstacle in using UV microscopy for the study of lignin distribution in different morphological regions of hardwoods is that hardwood lignins are composed of both guaiacyl and syringyl units and therefore the local ratio of these two also affects UV absorption [122)]. Light and electron microscopy of lignifying cells of wounded wheat leaves and the fungal colonization of the wounded tissues has been reported [155)]. Conventional electron microscopy has elucidated the fiber surface structure and fiber liberation in different pulping systems [194)].

The use of TEM and SEM has given valuable information about the changes in wood ultrastructure and about lignin modifications, as well as about the morphology of fungal attack on wood. Scanning electron microscopy has been coupled with an energy dispersive x-ray analyzer (EDXA) to determine the lignin distribution in wood cell walls [193]. The technique involves selective bromination of lignin in a non-aqueous system and the subsequent analysis of bromine concentrations in different morphological regions. The bromination of wood was shown to be specific for lignin [195] and bromine is stable under the electron beam bombardment. This procedure possibly will become a powerful tool, with reasonably high accuracy, for the localization and quantitative assay of lignin distribution in wood cells.

8.2.7 Chromatography

Gel permeation chromatography (GPC), high performance liquid chromatography (HPLC), and gas chromatography (GC) have been used for the identification of aromatic acids produced during degradation of lignin [102], changes in molecular weight distribution of Kraft lignin [107], lignosulfonates [28] and Björkman lignin [100] during fungal attack. Gel chromatography has been used to isolate and fractionate the lignin-carbohydrate complex [13] and to determine the molecular weight distribution of hydrothermally-degraded lignin [38]. A new fungal cultivation method (Petri dish technique) has been developed, which eases the study of molecular weight distribution by GPC [10].

Gas chromatography has been used for analyzing lignin derivative compounds [160], and hydrogenation products of lignin [207]. A method has been described for the characterization of lignins in untreated plant and geochemical samples containing as little as 10 mg of organic matter using gas capillary chromatography [92].

A well-known difficulty in the interpretation of a GPC curve is the lack of proper calibration standards. This difficulty is even more pronounced in cases of lignin. The possible difference in degree of association of lignin components in different solvents and the aggregation between the resulting complexes may seriously affect the calibration curve and elution profiles of lignin samples [202]. For example, it was shown that a change of elution profiles of lignin from bimodality to monomodality is possible by using dimethylformanide (DMF) or 0.1 M LiCl in DMF, respectively, as a solvent [39]. This effect, however, was not observed in experiments by Concin et al. [38]. Because of the polyelectrolytic nature of lignosulfonates, the choice of eluent may also affect the molecular weight distribution of this compound. Budin and Susa [28] have compared the fractionation of lignosulfonate on Sephadex gels, using water and electrolytes as eluent. They have concluded that the fractionation results of a certain lignin preparation may only be compared when these results have been obtained under similar conditions. Yan and Johnson [235], using the Flory Stockmayer distribution theory for condensation of polyfunctional polymers, have demonstrated that the initial chain size distribution is important for determining the shape of GPC elution curves, as well as for determining post gel properties.

8.3 Lignin Degrading Organisms

Nature has already developed efficient mechanisms for the modification and degradation of lignin. The biological degradation of lignin is one of the most important parts of the biospheric carbon-oxygen cycle. The efficiency of lignin decomposition in nature is evident from the lack of accumulation of lignin on a year-to-year basis. Efforts to find and isolated microorganisms with lignin-decomposing ability have already borne fruit, and the range of microorganisms known to attack lignin has been expanded. Although it is supposed that lignin degradation is the result of the cooperative action between different fungi, bacteria, and microflora in the soil, fungi still remain the most studied organisms of all the other inhabitants of the soil.

It should be noted that the term 'degradation' is used in a broad sense. It includes not only the complete decomposition to CO_2 and H_2O, but also various modifications, or bioalterations, such as demethylation and partial oxidation.

8.3.1 Fungi

Based solely on the types of decay they cause, fungi are classified as white-rot (basidiomycetes and a few ascomycetes), brown-rot (basidiomycetes), and soft-rot (ascomycetes) and fungi imperfecti. It is generally recognized that the complex polymers of lignin are first attacked by basidiomycetes, and are substantially degraded by some ascomycetes and imperfect fungi [119].

White-rot fungi attack unaltered lignin polymers, causing cleavage of interlignol bonds such as C_α-C_β, β-aryl ether, C_1-C_α, and aromatic ring cleavage (see Sect. 8.6). Oxidation of C_α and $C_\alpha=C_\beta$, aromatic hydroxylation, and demethylation of methoxyl groups is also possible [119]. Brown-rot fungi are mainly humifiers, causing only limited changes in lignin. These changes include demethylation of methoxyl groups, aromatic hydroxylation, and limited side-chain oxidation. The brown-rot fungi do not cleave lignin's aromatic rings efficiently or if they do open the rings, they are unable to make significant decomposition in the resulting lignin fragments [41]. Soft-rot fungi apparently degrade lignin-quite slowly and incompletely. Soft-rot decay is characterized by attacks on wood under moist conditions.

The highest lignin degradation rates have been reported for different types of basidiomycetes. For example, the lignin degradation in wheat straw by fungal colonists of ruminant dung was found to be sixfold greater with basidiomycetes than with ascomycetes [232]. A 65% reduction in lignin content (Klason method) in three weeks by *Pleurotus*, a white-rot sp. was observed in mushroom production on cotton straw [182]. Higher degradation rates were also reported under laboratory conditions [108,236]. Besides the type of fungus, the type of lignin also affects the rate of lignin degradation. In general, lignin from hardwoods (angiosperms) is more degradable by wood-rotting fungi than that from softwoods (gymnosperms). Lignin in hardwood (alder) was decomposed at least four times as rapidly as that in softwood (hemlock) [236]. Highley [94] has stated that the type of lignin is more a factor in the slow decay rate of softwoods than the amount of lignin. The molecular basis for this relative resistance of softwoods is not known. Comparisons between the lignin degradability of wood-rotting and grass-rotting fungi should also be made. Recently Antai and Crawford [12] have reported that *Coriolus versicolor*, a white-rot fungus, and *Poria placenta*, a brown-rot fungus, can decompose grass lignin more rapidly than hardwood (maple) or softwood (spruce) lignin.

Eucaryotic microorganisms other than the three major groups mentioned above have not yet been closely examid for ligninolytic activity. Recently, Sutherland and

Crawford [217] have reported the decomposition of ^{14}C-labeled maple and spruce lignin by several marine and estuarine fungi. Some cultures accumulated substantial ^{14}C in the medium, indicating a high rate of lignin solubilization rather than complete degradation to $^{14}CO_2$. Studies on lignin degradability by yeast are also inadequate and must be confirmed. A species of *Candida* isolated from decaying leafy material was reported to have degraded the Kraft lignin in a bioreactor after 36 h and produced, as verified by elution profiles of lignin, aromatic intermediates [36]. However, the analysis was made with a culture supernatant under pH values optimum for yeast growth in which Kraft lignin could only be partially solubilized. Furthermore, the possible interference of cell metabolites on the elution pattern of lignin was not discussed.

There is also a report stating that the labeled dioxane lignin from corn is degraded to CO_2 (4%) by yeast *Trichosporin fermentans* [4]. The results suggest further that the major action of this yeast on lignin is in depolymerization so that significant degradation of the aromatic ring does not occur. This is promising regarding the production of aromatics for use as liquid fuel.

8.3.2 Bacteria

Bacteria probably play a secondary role in lignin degradation after fungi in lignin degradation in various soil and water environments. Most of the literature on bacterial degradation of lignin deals with the decomposition of monomeric and dimeric compounds such as methoxylated aromatic acids [132]. Only a few strains have been shown to be capable of attacking lignin and lignin derivatives. The most active strains are those from actinomycetes, such as *Streptomycete* and *Nocardia* spp. [41].

Several noncellulolytic gram-negative aerobic bacteria have been reported to be capable of degrading dioxane and milled wood poplar lignins at rates ranging between 4% to 20% within a 7 day period (estimated on the basis of A_{280} of residual lignin in dioxane-water) [174]. These strains were identified as *Pseudomonas*, *Xanthomonas*, and *Acinetobacter*. They could also degrade poplar wood in situ (40%–57%, on the basis of A_{280} of residual lignin solubilized in acetyl bromide/acetic acid) [173]. In situ lignin degradation was assayed for by monitoring $^{14}CO_2$ evolution from cultures containing [^{14}C] lignin poplar lignocellulose using *Pseudomonas* sp. [173]. The total $^{14}CO_2$ recovery in 23 amounted to 1–4% and depended on the degree of wood milling. Electron microscopic investigations showed that both the primary and the secondary walls of poplar wood could be delignified by *Pseudomonas* [157].

Two strains of *Bacillus polymyxa* caused a loss of 42% in lignin content of Scots pine sapwood (acetyl bromide method) [205]. However, UV-microspectrophotometry of bacterial-inoculated cell walls gave no indication of any changes in the remaining lignin. Crawford [40] and Antai [11] showed that *Streptomyces* could degrade grass and corn lignin more extensively than hardwood (maple) or softwood (spruce) lignins. After 12 weeks, 39–44% of the grass lignin could be depleted (Klason method) [11].

Bacteria have also been investigated for use in the degradation of industrial lignins. Chlorinated lignin from bleachery discharges, which is known to be resistant to biodegradation, was used, but no bacterial strain could be found to depolymerize the high molecular weight of the lignin [165]. However, biodegradation of 2,4-

carbon source has also been degraded, by bacteria. A strain of *Streptomyces* degraded MWL in a mineral medium [17]. Degradation of poplar dioxane lignin without carbon supplementation has also been reported for different gram-negative aerobic bacteria [174].

Nitrogen: The inhibitory effect of the nutrient nitrogen above growth-limiting concentrations has been shown for *Phanerochaete chrysosporium*. The lignin degradation takes place only under nitrogen starvation conditions [60,118,187,236]. The extent of inhibition by nitrogen is not the same for different nitrogeneous compounds. Glutamate, glutamine, and histidine were the most effective inhibitors of ligninolytic activity (Table 13) [65,210]. The oxidation of the lignin-related phenol, 4-hydroxy-3-methoxyacetophenone, responded similarly to the addition of nitrogen [65]. β-guaiacyl ether-linked lignin dimeric compounds were also cleaved only at low nitrogen levels [57,228]. The onset and suppression of ligninolytic activity do not represent changes in glucose catabolite repression in response to nitrogen starvation [65]. The addition of NH_4^+ and L-glutamate to ligninolytic cultures of *P. chrysosporium* increased the specific activities of glutamine synthetase and NADP- and NAD-glutamate dehydrogenases [63]. Therefore, the repression of ligninolytic activity by the nutrient nitrogen was interpreted as the biochemical repression of enzymes associated with lingin degradation. Kirk [119] has hypothesized that glutamate metabolism might play a role in the regulation of lignin degradation as a part of the secondary metabolism in *P. chrysosporium*. Kirk et al. are also studying the nature of repression of the ligninolytic system caused by the addition of nitrogen.

However, generalizations concerning the nitrogen effect should not be made. Zadrazil and Brunnert [238] have investigated the influence of nitrogen supplementation on degradation of straw-lignin in solid cultivation by various higher fungi. They found that the ability to degrade lignin was influenced differently by nitrogen supplementation. The lignin decomposition abilities of some of the tested fungi were depressed, for one fungus was not correlated, and for another fungus, nitrogen supplementation even stimulated lignin degradation. Concerning bacteria, the effect of nitrogen supplementation on the degradation of lignin by *Streptomyces bodius* was examined by Barder and Crawford [17]. Unlike *P. chrysosporium*, an enhancement of lignin degradation was achieved at high levels of organic nitrogen. High concentrations of inorganic nitrogen were inhibitory. The inhibition was assumed to be due to specific ion effects on polymer degradation, rather than due to salt concentration effects on cellular growth [17].

Other culture constituents: Not much data are available about the effects of other medium ingredients on the ligninolytic activity of microorganisms. According to Jeffries et al. [109], limitation of carbohydrates or sulfur, and not limitation of phosphorus, can trigger ligninolytic activity of *P. chrysosporium*. In terms of sulfur limitation, contrary results have been observed by Reid [187]. However, nitrogen seems to be the only element found so far that has a predominant effect on the regulation of lignin degradation since sulfate limitation does not induce high rates of lignin degradation in the presence of high nitrogen levels [187].

The influence of limitation in concentrations of some trace elements has also been studied, although not adequately. Some improvements in the initial rate of lignin degradation could be achieved by limiting Zn^{2+}, Fe^{2+}, and Mo^{6+}. It has been reported that high concentrations of Mn^{2+} were inhibitory and that the balance of Mg^{2+} and Ca^{2+} were important [109]. Norris [169] has reported that the degradation of DHPs by *Fusarium solani* is drastically reduced when D_2O is the liquid medium.

Culture conditions: Culture conditions such as temperature and pH that favor lignin degradation have not been adequately studied. The optimum conditions for growth and lignin degradation may not be the same. Drew and Kadam [51] have found that in a period of 15 days, about 4 times more ^{14}C-Kraft lignin can be converted to CO_2 by *P. chrysosporium* at 28 C than at 38 C, although the latter temperature allows greater growth. The optimum pH for lignin degradation by this fungus is close to pH 4, whereas the optimum for growth seems to be pH values close to 5 [120].

Buffering of the culture medium is apparently also important. Initial buffering of the culture medium to pH 4.0 or 5.0 has significantly increased degradation by *F. solani* of lignin and several aromatic acids [169]. The degradation rate of ^{14}C-lignins to $^{14}CO_2$ by *P. chrysosporium* in cultures buffered with 2,2-dimethylsuccinate was twice that of *o*-phthalate buffered cultures [64], which could be accounted for by *o*-phthalate inhibition of the activity of the ligninolytic system. Nakatsubo et al. [162] have used polyacrylic acid, which does not interfere with extraction of degradation products.

Oxygen: Chemical and spectroscopic studies of decayed lignin have shown that lignin degradation is largely oxidative. Yang et al. [236] have found that the rate of degradation of lignin in alder pulp by *P. chrysosporium* is at least 50% more rapid in cultures under 100% oxygen than in those under air. The metabolism of β-guaiacyl ether-linked lignin dimeric compounds was also greater under pure oxygen than air [228]. Bar-Lev and Kirk [16] have shown that the positive effect of O_2 concentration appears first after primary growth, and they concluded that oxygen itself may not be responsible for the induction of the ligninolytic system but that its elevated concentration enhances the rate of lignin metabolism after the system is formed. Increasing oxygen partial pressure above 1 atm, however, did not improve lignin degradation. Oxygen pressures higher than 2 atm first inhibited growth and then killed the fungus due to increased oxygen concentration [188]. Enhancement of lignin metabolism because of increased oxygen pressure was also observed for several other white-rot fungi such as *Coriolus versicolor* [7] [p. 88], [189], *Pycnoporus cinnabarinus, Lentinus edodes, Grifola frondosa, Polyporus brumalis, Merulius tremellosus* [189] and imperfect fungus *Fusarium solani* [169]. However, among the tested white-rot fungi, some species such as *Gloeporus dichrous, Pleurotus ostreatus*, and *Bondarzewia berkeleyi* were less responsive to increased oxygen pressure [189]. It then seems possible that different species exhibit different responses in this respect. Data on the possible effect of oxygen on bacterial degradation of lignin is still incomplete.

Although the oxidative nature of lignin degradation has been emphasized, some reductive reactions could also take place. The formation of primary and secondary alcohol groups and of methylene groups has been observed which are partly caused by reductive processes [9,14,55,79]. Zeikus [241] has postulated that oxygen is required for the biological mediated depolymerization of lignin but not for the microbial decomposition of model lignins that contain ether linkages or for the soluble aromatic compounds that contain an oxygen function in the molecule.

Lignin probably cannot be degraded in a fully anaerobic environment [240]. Whether any modification in the lignin structure could take place in the absence of oxygen is not yet known. Although anaerobic bacteria are not considered to be lignin utilizers, Akin [2] could isolate a filamentous facultative anaerobic bacterium from rumen fluid that could attack lignified tissues in forage grasses under anaerobiosis. The attack was shown by scanning electron microscopy, and lignin content was not measured.

Whereas there are few reports on lignin degradation under anaerobic conditions, evidence can be found regarding the anaerobic degradation of lignin-derived compounds [14,37,91]. Investigations have shown that benzoic compounds with appropriate patterns of methoxyl group substitutions, such as those found in 3,4,5-trimethoxybenzoic, syringic, and 3-O-methylgallic acids, give rise to methanol when subjected to bacterial degradation [49]. Kaiser and Hanselmann [111] have reported that 3,4 disubstitutions of lignin-derived monomers were converted to catechol under anaerobic conditions, whereas 3,4,5-trisubstitutions of these monomers were mineralized to CH_4 and CO_2 under the same conditions. The mechanisms of anaerobic degradation of aromatic compounds and the question of whether aromatic derivatives can be used as a source of methane requires more investigation.

Mode of cultivation: A strong inhibition of lignin degradation from agitation has been observed in cultures of *Phanerochaete chrysosporium* [236]. Degradation of lignin in shallow, liquid-phase, stationary cultures was 10 times as rapid as in agitated cultures. It is postulated that this effect is due to pellet formation as the result of agitation. When fungi are grown as mats, agitation does not affect the lignin degradation. The metabolism of β-guaiacyl ether-linked lignin dimeric compounds [228] and dimethoxycinnamyl alcohol [210] by *P. chrysosporium* were also greater in stationary cultures than in shaking cultures. Similar investigations must also be carried out for other lignin-degrading organisms.

Some investigators have used solid or semi-solid cultivation modes [11,12,182,238] which are especially useful for improving the digestibility of lignocellulosic material and the production of edible mushrooms. Not much work has been conducted yet to assess the novel bioprocesses for lignin degradation. Amer [7] [p. 94] has compared the lignin degradation of *Coriolus versicolor* in solid state and fluidized bed systems using wood chips and has concluded that the solid state process leads to greater lignin degradation. The fluidized bed process favored the degradation of carbohydrates rather than the degradation of lignin. Novel operation systems should be developed to improve lignin degradation in whole wood. Recently, evidence has shown that by choosing the optimum ratio of surface to volume of the culture, the rate and extent of lignin degradation can be improved [87].

8.5 Enzymology of Lignin Degradation

The question whether lignin decomposition by microorganisms is an enzymic oriented event, has engaged the attention of many investigators. The postulation that lignin degradation has an oxidative nature has made investigators consider the possible involvement of oxidizing enzymes in lignin degradation. The enzymes in question are aromatic alcohol dehydrogenases, phenol oxidases (laccase, peroxidase, tyrosinase), mono- and di-oxygenases [97]. Oxidation reactions catalyzed by the involved enzymes are supposed to be nonspecific and nonstereoselective [115,162,168].

Phenol-oxidizing enzymes have been studied the most, but their participation in lignin degradation is still a matter of controversy. There are indications that the cleavage of alkylphenyl C-C bonds [76,112,116] and α, β C-C bonds [56] can be accomplished by a phenoloxidase-catalyzed reaction. A laccase preparation from *Coriolus versicolor* could remove hydrogen from the α and β positions of the side chain of DHP [168]. Weinstein et al. [228] doubt however, that the cleavage of alkylphenyl C-C bond in syringylglycol-β-guaiacyl can be catalized by a phenol oxidase enzyme. Transformation reactions of phenylcoumaran compounds in cultures of *P. chrysosporium* were presumed to be enzyme mediated [220].

Several lignin-degrading *Fusarium* spp. have been reported to excrete an enzyme which catalyzes the oxidation of the α,β-unsaturated primary alcohol group in the side chain of lignin-related aromatic compounds such as coniferyl alcohol, cinnamyl alcohol, dehydrodiconiferyl alcohol, and guaiacylglycerol-β-coniferyl ether to corresponding aldehydes [104]. The enzyme seemed to be an aromatic alcohol oxidase, and it was not clear whether the oxidation takes place due to one or several enzymes. In a further study using d,l-syringaresinol, a β-β′ linked lignin model, an intracellular enzyme was isolated from *Fusarium* sp. that oxidized phenols such as guaiacol, and it was assumed that the enzyme was a laccase type [103,112]. An intracellular enzyme from *Fusarium* that seemed to be different from laccase or peroxidase was assumed to be responsible for the oxidation of α-carbons of pinoresinol [102]. Both laccase and peroxidase were found in association with the cells of yeast *Trichosporon fermentans*, which were reported to have been able to degrade lignosulfonate [4], but the correlation between these enzymes and lignin degradability was not reported. Vanillate hydrolase, an intracellular enzyme, was isolated from *Sporotrichum pulverulentum* (= *P. chrysosporium*) and was declared to participate in the oxidative decarboxylation of vanillate to methoxyhydroquinone [9,29]. This enzyme was also found in many brown-rot and white-rot fungi, but not in soft-rot fungi [29].

Kaplan [113] examined the reactivity of different phenol oxidases (laccase, peroidase, tyrosinase) with lignins and lignin model compounds but did not find any evidence for the production of stable, low-molecular weight compounds. Haars and Hüttermann [80] studied the function of laccase in the white-rot fungus *Fomes annosus* and concluded that the lignosulfonate is degraded to the same extent, regardless of whether laccase is active or not. It was proposed that the enzyme cellobiose quinone oxidoreductase (CBQase) is linked to the degradation of cellulose and lignin, but further studies have cast doubt on the role previously assigned to CBQase in lignin degradation [4], [p. 25].

The role of laccase has been discussed not only in connection with lignin depolymerization, but also with polymerization. The molecular weight of white-rotted lignin is usually higher than sound lignin (Table 11), indicating an increase in the degree of lignin polymerization. The fact that lignin degradation products of low molecular weight have always been found only in small quantities may also indicate the repolymerization of such products. Hüttermann et al. [81,100] have shown that laccase is responsible for the observed partial polymerization of lignosulfonate, of Kraft lignin and of Björkman spruce lignin that takes place in the culture medium of *F. annosus*. Polymerization could be prevented by the addition of laccase inhibitors such as thioglycolic acid or sodium oxide to the culture medium.

Oxidative polymerization of the lignin monomer, coniferyl alcohol, also occurred in the presence but not in the absence of laccase activity [80]. Extracellular laccase from the fungus *Rhizoctonia praticola* could catalyze the formation of various oligomers ranging from dimers to hexamers of vanillic acid, orcinol, and vanillin [23,24], and from syringic acid [138]. The oxidative coupling of vanillic acid was also catalyzed by peroxidase but not by tyrosinase [24]. A soil-extracted phenoloxidase-like enzyme produced comparable results [214] and could catalyze the conversion of different phenolic compounds to oligomeric products by C—C and C—O coupling reactions.

The correlation between enzymic activity, ligninolytic activity and physiological parameters such as oxygen concentration, nitrogen limitation, mode of cultivation, etc. that are known to induce ligninolytic systems has not yet been well documented. Keyser et al. [118] reported that the addition of nitrogen to cultures immediately prior to the usual time of the ligninolytic activity delayed its appearance, and they suggested that nitrogen interferred with the synthesis of the enzyme system. Goldsby et al. [76] unexpectedly observed that the alkylphenyl bond in guaiacylglycol- and guaiacylglycerol-β-guaiacyl ether was cleaved at more significant rates in a medium with high nitrogen content. They concluded that perhaps only a fraction of the enzymes involved in the complete degradation of lignin and lignin model compounds are under nitrogen control. How enzyme reactions of lignin degradation or polymerization would respond to different lignin concentrations, medium ingredients, oxygen concentrations, and other environmental conditions are unknown.

The enzymology of lignin biodegradation, as understood from the uncertainties and partial discrepancies between different investigators, remains largely obscure, despite the substantial efforts that have been made. Difficulties in the purification and isolation of enzymes and the problems involved in the quantitative analysis of the functional groups formed by the chemical modification of complex lignin polymers with nonspecific enzymes are the main barriers to progress in understanding the enzymology of lignin degradation. Enzyme isolation techniques could cause inactivation of one or more components of the system. Enzyme studies have been made mostly using lignin model compounds. The enzymes involved in the degradaton of such compounds are typically intracellular and highly substrate-specific, and it is unlikely that they can attack an extracellular complex polymer such as lignin [41]. Oxidative reactions catalyzed by oxygenases require coenzymes such as NAD(P)H, and the excretion of such coenzymes by fungi is rather unlikely. Therefore, the occurrence of initial degradation reactions of lignin at an enzymic level could be doubted. Phenol oxydases are certainly at least partially important in lignin decomposition by white-rot fungi. However, the phenol oxydases have not been purified and shown definitely capable of depolymerizing lignin or even of releasing low molecular weight fragments in significant amounts. If these enzymes are only partially involved in lignin breakdown, what is then their main function? Some authors have suggested that these enzymes may only have a detoxifying function. However, Haars and Hüttermann [80] doubted the detoxifying role of phenol oxidases. They showed that catechol and guaiacol were converted to substances toxic to fungi in the presence of laccase.

A working hypothesis by Hall [85] suggested that perhaps the enzymes responsible for lignin degradation do not interact with the polymer itself but generate reactive diffusible species such as superoxide radical anions which attack the polymer. In fact, the mechanisms of enzymatically catalyzed oxidative degradation of lignins proposed thus far are similar to, if not identical with, those suggested for purely chemical oxidative degradation [75]. Testing this hypothesis requires that specific degradation reactions be identified, but until recently there was no direct evidence to support these speculations. The possible involvement of singlet oxygen ($^1O_2$) in the degradation of lignin by *Phanerochaete chrysosporium* was recently examined by Nakatsubo et al. [161]. The singlet oxygen was formed photochemically and labeled 1,2-bis(3-methoxy-4-alkoxyphenyl)propane-1,3 diol was used as the lignin model. It could be shown that the singlet oxygen played an integral role in lignin degradation, and the chemistry of lignin polymer oxidation by *P. chrysosporium* was similar to the chemistry of lignin oxidation by $^1O_2$. Amer [7] [p. 71] found the extracellular superoxide radical (O_2^-) and hydrogen peroxide in the ligninolytic cultures of *Coriolus versicolor*. The production of these reduced oxygen species were induced under conditions similar to those necessary for the induction of ligninolytic activity, i.e., nitrogen limitation and standing cultivaton under an atmosphere of oxygen. These similarities suggest that reductive oxygen species may be responsible for the initial breakdown of the lignin macromolecule.

8.6 Mechanisms of Lignin Degradation

The question why fungi degrade lignin at all remains unanswered. Lignin apparently does not serve as a growth substrate [119]. Statements such as 'fungi must degrade lignin to gain access to the polysaccharides of the wood' or 'it degrades lignin for survival but not for growth' [241] do not prove satisfactory, because lignin can be degraded in a submerged culture while a simply assimilable and soluble carbon and energy source such as glucose is still present in the culture [108,118].

An interesting question concerns the patterns of the lignin-degrading system. Where is the zone of influence of this system? Can the lignin-degrading system diffuse and move from one point to the other point within the media? If direct contact between the producer of the 'lignin-degrading system', e.g. fungal hyphae, and lignin is necessary for the system to become induced or operative, then it would become clear why attempts to detect cell-free activity have not been successful to date. Electron microscopic examination of decayed wood has shown that the lignin-degrading system diffuses from fungal hyphae and that lignin degradation does not require true contact between the fungal cell wall and lignin [192]. A contrary conclusion was drawn by Rosenberg [190]. A diffusion chamber consisting of two aerobic compartments separated by a bacteriological membrane filter was used to study the patterns of diffusibility of lignin-degrading systems for a number of known lignin-degrading fungi. Only *Coriolus versicolor* and a strain of *Trichoderma reesei* displayed slight (ca. 14%) diffusible lignin-degrading activity. None of the other organisms tested gave evidence of being able to produce a diffusible lignin-degrading system. The conclusion was that the lignin-degrading systems may be nondiffusible.

Whether lignin is degraded via low molecular weight aromatic products or by the cleavage of the aromatic nuclei still bound to the polymer is not yet clear. The possible cleavage of the aromatic ring still bound to the polymer is supported in Refs. [35,55,108]. Chen et al. [32] support the idea that the primary mode of lignin degradation is in part via low molecular aromatic acids. Haars and Hüttermann [81] have indicated that the basidomycete *Fomes annosus* is unable to split the lignin molecule directly into intermediate-sized pieces and that lignin biodegradation in vivo follows the mechanisms of exo-degradation. It is, however, likely that both mechanisms are involved and that lignin degradation proceeds by a mixed function, that is, ring-fusion reactions in a largely intact polymer, accompanied by the release of some low molecular weight lignin fragments.

Studies for the elucidaiton of initial steps in lignin degradation have mainly been made using white-rot fungi and there has been little effort to show the degradation patterns that exist within other microbial groups. Although different bateria genera are now known to contain lignin-decomposing species, the mode of attack is still unknown. Crawford et al. [42] have postulated that lignin degradation by the actinomycete *Streptomyces viridosporus* probably proceeds by the mechanisms of white-rot decay. Björkman milled softwood lignin was oxidized by this bacteria, as revealed by degradative chemical and elemental analysis, in a similar way to white-rot decay, which means that lignin carbon content decreased, oxygen content increased, the 4-hydroxy-3-methoxyphenyl-propane subunit structure was lost, demethylation took place, and the remaining lignin was polymeric but was significantly dearomatized. The only exception was an increase in hydrogen content, which could be attributed to the saturation of branched C—C bonds resistant to further bacterial attack [42].

Studies have been made on the sequential pathways after primary attack and depolymerization using lignin model compounds. Work with low molecular lignin models, however, has provided only limited information on possible pathways. The most recent of the suggested degradation pathways are listed in Table 10. Suggestions have been made concerning the requirements and necessary conditions under which certain reactions of lignin breakdown can take place.

Ander et al. [9] have shown that decarboxylation of vanillic acid by white-rot fungi occurs prior to ring cleavage and that cleavage of the benzene nucleus by fungi requires hydroxyl substitution in the benzene ring. However, such a cleavage has been observed in benzene nucleus lacking hydroxyl substitutions [169]. Concerning the cleavage of C_α—C_β bond, Weinstein et al. [228] have demonstrated that neither the free phenolic functions on the ring nor ring cleavage is a prerequisite for the cleavage of the C_α—C_β bond, as had been postulated earlier. Rast et al. [186] have studied the influence of ring substituents on the degradation of veratrylglycerol-β-guaiacyl ether by bacteria. The conclusion was that the first step in decomposition is strongly influenced by the kind and position of substituents on the phenyl ether moiety, but it is doubtful whether gram-negative and gram-positive bacteria would behave in the same manner.

Considering the discussions and conclusions concerning the conditions that could lead to cleavage of certain bond or to the addition or elimination of functions and groups, etc., there is still little evidence to show how degradation of a single aromatic ring or a lignin model compound and the degradation of a lignin polymer is correlated and it is doubtful whether these cleavages lead to any net degradation of lignin.

8.7 The Potential for Biological Lignin Degradation

The ability of microorganisms to degrade lignin has been used in several applied and industrial projects. These projects have become topical mainly because of expected shortages in the world's energy and food supplies, the need to find ways to utilize waste, and many environmental problems which have been caused by man and industry. Examples of potential uses for microbial lignin degradation are the biological treatment of lignocellulosic materials, biological pulping, and effluent treatment.

The encrustation of cellulose with lignin makes lignocellulose an unsuitable substrate for enzyme production, as well as for enzymic hydrolysis. Lignin inhibits cell wall digestion by ruminants through various mechanisms [166]. The pretreatment of lignocellulose necessary to accelerate enzymatic attack is similar to the treatment needed to increase the digestability of food by ruminants. The biological removal of lignin could possibly offer a low cost route to enhancing the digestability of lignocellulose. Nutritional evaluation studies performed on microbial biomass or feeds enriched with microbes have, in some cases, demonstrated excellent results [158]. Parameters affecting the ability of white-rot fungi to improve digestability of straw have been investigated [239]. However, data comparing the costs of biological treatments to those of non-biological processes is still needed [6] [p. 103].

It has been reported that enough lignin can be degraded to cause a decrease in the amount of energy required for the production of thermomechanical pulp if wood chips are pretreated by cellulaseless mutants of white-rot fungi [59,60]. Unbleached Kraft pulp may be partially delignified during incubation with ligninolytic fungi, which would reduce the amount of chemicals necessary for bleaching. Why this method has not yet been employed by the paper industries is not clear.

Another area of applied research has been in the fungal decolorization of Kraft bleach plant effluents. Lignin degradation products and chlorinated lignins are the main contributors to the color and toxicity of bleach plant effluents. Pulp and paper mill effluents, particularly those produced from the chemical pulping of softwood, are dark brown in color and their discharge into receiving waters causes serious pollution problems [222]. The removal of lignin from these waste waters is not only of great importance from the ecological standpoint, but is also highly desirable for cleaning up process waters in closed paper mills. The effluent volume and load can, of course, be reduced using processes such as ultrafiltration, carbon adsorption, lime treatment as well as oxygen bleaching and dynamic bleaching. Furthermore, lignin can be removed from spent liquors by the addition of chemicals [137]. Coagulation takes place through interparticle bridging by a non-ionic polymer, such as polyethylene oxide in the presence of an electrolyte. However, these processes and technologies, first of all, are expensive, and secondly, do not offer concrete solutions to the problem of waste disposal. Therefore, no acceptable procedure has been found that effectively eliminates the color and toxicity of these effluents.

However, recent research has shown that fungi such as *P. chrysosporium* can decolorize Kraft bleach pulp effluent [53,215]. The chromophore-bearing material was isolated and characterized. A comparison of the elemental analysis, color yield, molecular size distribution of the original effluent and the partially decolorized bleach pulp effluent showed that the decolorization was the result of simultaneous chromophore destruction and polymer decompositon [215]. In this procedure, paper mill waste primary sludge may be used as a growth substrate for fungi. More than 80% of the color (per day) has been removed and fungal mats, once grown, will sustain the same rate of decolorization for up to 60 days [19]. The possibility of microbial desulfonation of lignosulfonate has also been reported [211]. In this experiment, 14 isolates were screened, two of which could release over 60% of the sulfate ester present in the lignosulfonate during growth. This would present an alternative to chemical and physical desulfonation processes, which are known to be partial and nonspecific.

9 Outlook

The knowledge of chemistry gained from one hundred years history of pulping, fifty years experience with the dehydrogenation theory, and the present modern strategies for the synthesis of oligolignols have provided many possibilities for the future of lignin chemistry regarding the complete utilization of lignins. Enormous efforts are being made worldwide in both the chemical and biological aspects to approach this goal. The results obtained thus far reveal many possibilities for the complete utilization of this gift of nature, lignin. Some lignin research projects have already reached the stage of realization. Others, which have not, either because of inadequate investigations or for economical reasons, will certainly be realized in the next century.

Among the possibilities offered by chemical processes to date, the primary uses for isolated lignin products and lignosulfonates have been macromolecular ones, such as dispersants or emulsion stabilizers in solution systems, as well as in lignin-based plywood, particleboard adhesives, and asphalt binders. However, the quantities used now for these applications are only a fraction of the total amount of lignin available from the waste liquors of pulping processes. One possibility for a large-volume, economically acceptable way of utilizing lignins for purposes other than energy generation seems to be in the field of enchanced oil recovery. In this area, displacement tests using mixtures of lignosulfonates and petroleum sulfonates have produced very exciting results.

Another possible use, lignin fragmentation and chemical conversion to different phenolic products, is still highly dependent on lignin's economic competitiveness with petroleum-derived products. Most phenolic products can still be made more cheaply from petroleum than from wood. Limitations in conversion efficiency and product recovery costs are still barriers to the realization of these products. Therefore, prospects for the production of phenols from lignin will depend upon the availability of petroleum in the future. A small market volume and relatively low product volume are further limiting factors to the utilization of lignin for special chemicals. In this respect, it may be preferable that research be directed towards the production of lignochemicals which have structures that cannot be obtained from petroleum. The conversion of lignin to useful products should take more advantage of lignin's existing polymeric structure and production of macromolecules, which can be reassembled into useful polymers.

Research on the role that microorganisms can play in lignin transformation and utilization is rather new, and it is too early to speculate on the commercial significance of lignin biotransformation. Considering the fact that lignin is one of the most resistant compounds to biological degradation, the discovery of microorganisms able to degrade lignin and lignin compounds and the partial understanding of the physiology and biochemistry of this phenomenon should be regarded as a significant contribution to the field of lignin research. Current research shows promise for future developments in bioconversion. Progress is being made towards understanding and possibly utilizing the process of microbial biodegradation of lignin. Microorganisms can be used to convert or destroy plant materials or residues or industrial by-products derived from lignocellulose conversion processes.

One project that could soon become realized is the microbial decolorization of Kraft bleach plant effluents and the removal of lignin from waste waters. The processing of lignin-containing plant wastes remains a major goal of biotechnology in energy production. Other contributions to lignin degradation are in biomass conversion. There is increasing interest in the development of bioconversion processes in which waste lignocellulose material can be processed into sugar, alcohol, and other organic solvents or protein. In these processes, lignin is perhaps the most serious impediment to the development of a successful bioconversion process. An economical method of converting lignocellulose to other useful materials requires the integrated use of all three components of lignocellulose. Improvements in fungal selectivity towards lignin by mutation, genetic manipulation, or nutrient modification should help to reduce energy consumption and the cost of pretreatment of lignocellulose materials.

The enzymes participating directly in the metabolism of lignin have yet to be identified. Since no low molecular weight products appear to accumulate during lignin biodegradation in significant amounts, applications other than the production of chemicals from lignin biodegradation should be investigated. The most attractive route towards bioprocessing would be to use microorganisms for the modification of lignin polymers. Lignins recovered from pulping processes have limited polymeric properties and are quite limited in chemical functionality. Demethylation, hydroxylation, side chain shortening, and ring cleavage of lignin could satisfactorily alter polymeric lignins for chemical modification. Chemical modifications to improve functionality are energy and materials resource intensive. Specific modifications of lignin through the action of enzymes might result in a more reactive residual unit as a building block for polymer applications.

Lignin research has now been stimulated primarily by hopes for eventual industrial application, but it is still too early for a proper cost-benefit analysis of most of the proposed biological processes. A better understanding of the biodegradation of lignin can be put to use in developing new and improved preservatives. Research in this area will also have consequences for other bioprocesses such as microbial degradation of herbicides and fungicides.

10 Notes added in Proof

Since the manuscript has been prepared some significant results have been reported.

Lignin Pyrolysis, Reaction Pathway

Reaction pathway in pyrolysis of phenethyl phenyl ether (PPE), a model of the beta-ether linkages prevalent in lignin, has been studied. The results permit interpretation of the thermal stabilities of lignin related beta-ethers: Klein, M. T., Virk, P. S.: Ind. Eng. Chem. Fundam. *22*, 35 (1983)

Lignin Analysis and Identification

The effect of metal ions on the UV spectra of humic acid, tannic acid, and lignosulfonic acid in natural waters has been investigated:
Alberts, J. J.: Water Res. *16*, 1273 (1982)

Comparison of various lignin assays such as gravimetric and acetyl bromide for determining ruminal digestion of roughages by lambs has demonstrated that the choice of analytical method and extent of recovery may drastically affect the interpretation of digesta flow measurements calculated by reference to lignin:
Muntifering, R. B.: J. Anim. Sci. *55*, 432 (1982)

Spent bleach liquor from a pine Kraft mill has been analyzed for chlorinated guaiacols. 9 chlorinated guaiacols were found in concentrations between 0.004 to 1.57 ppm and their structures were confirmed by combined GS-MS. The highest and the lowest concentration belonged to 3,4,5-trichloroguaiacol and 3,4,6-trichloroguaiacol respectively:
Knuutinen, J.: J. Chromatogr. *248*, 289 (1982)

Anaerobic Degradation of Lignin

Oligolignols of 300–1400 molecular weight may be degraded anaerobically by a mixed bacterial population to yield low-molecular weight soluble compounds and end products of methanogenesis:
Colberg, P. J., Young, L. Y.: Abstr. Ann. Meet. Am. Soc. Microbiol. 1982, 199 (1982)

Labeled synthetic lignins could not be converted to CO_2 or methane in anaerobic environments of anoxic sediments, but the chemically modified, partially soluble synthetic lignins were degraded. Sulfate addition caused sulfidogenesis and limited methanogenesis. Low-molecular weight lignins were also degraded but high-molecular weight lignins were not:
Zeikus, J. G., Wellstein, A. L., Kirk, T. K.: FEMS Microbiol. Lett. *15*, 193 (1982)

Effect of Low Molecular Lignin Fragments on Lignin Biodegradation

Low-molecular weight ($MW_{ave.} = 300$), water-soluble lignin derivates of Kraft pine lignin inhibit lignin solubilization as well as lignin degradation by *Phanerochaete chrysosporium*:
Haltmeier, T.: Biokonversion von Lignin und Polysacchariden aus alkalisch behandelter Lignocellulose. Thesis, ETH Zürich 1983

Effect of Nitrogen on Lignin and Cellulose Degradation by Basidiomycetes

Increasing concentration of nitrogen in the culture medium has yielded higher CO_2 evolution from the lignocellulosics by three basidiomycetes. Lignin degradation by *Pleurotus ostreatus* and *P. chrysosporium* was suppressed at increased levels of N, for the third fungus, a coprophilus, the ligninolytic activity remained unaffected:
Freer, S. N., Detroy, R. W.: Mycologia *74*, 943 (1982)

Improvement in Lignin Biodegradation

Increasing the ratio of surface to volume of the standing cultures of *P. chrysosporium* has resulted significant improvement in the rate and extent of the lignin degradation. The high-molecular weight lignin, at concentrations of $1\,g\,l^{-1}$, was degraded completely to CO_2 (60–70%) and low-molecular weight, water-soluble compounds (30–40%) within an active ligninolytic period of 2–3 days. Different lignins including Kraft pine lignin, alkali lignin, dioxane lignin and lignosulfonate have resulted the same high degradation rates:
Leisola, M. et al.: Eur. J. Appl. Microbiol. Biotech. (in press 1983)
Leisola, M., Ulmer, D., Fiechter, A.: Eur. J. Appl. Microbiol. Biotech. (in press 1983)
Ulmer, D. et al.: Appl. Environ. Microbiol. (in press 1983)

Involvement of Hydroxyl Radicals in Lignin Degradation

The hydrogen peroxide production in ligninolytic *Phanerochaete chrysosporium* cells has been localized. H_2O_2 production appeared to be localized in the periplasmic space of the cells from ligninolytic cultures grown for 14 days in nitrogen-limited medium. Cells with little ligninolytic activity had also low specific activity for H_2O_2 production:
Forney, L. J., Reddy, C. A., Pankratz, H. S.: Appl. Environ. Microbiol. *44*, 732 (1982)

A variety of hydroxyl radical scavengers including thiourea, benzoic acid, mannitol, catalase, salicylate and 4-o-methylisoeugenol had an inhibitory effect on ligninolytic ability of *P. chrysosporium*:
Kutsuki, H., Gold, M. H.: Biochem. Biophys. Res. Commun. *109*, 320 (1982)
Forney, L. J. et al.: J. Biol. Chem. *257*, 11455 (1982)

Lignin Biodegradation Products

Absorbance spectra and gel chromatography of the aspen lignin after fungal attack have shown formation of water-soluble primary lignin fragments before further metabolization to CO_2:
Reid, I. D., Abrams, G. D., Pepper, J. M.: Can. J. Bot. *60*, 2357 (1982)

Cellobiose Quinone Oxidoreductase (CBQ)

Factors affecting the CBQ-production have been found to be cellulose or cellobiose as inducers, culture temperature of 37 C, standing cultures and oxygen. Under optimal conditions CBQ levels could be increased by 10 fold to $0.0222\,U\,ml^{-1}$ in 5 days in comparison to prior yields of $0.002\,U\,ml^{-1}$:

Kelleher, T., Jeffries, T. W.: Abstr. Ann. Meet. Am. Soc. Microbiol. 1982, 138 (1982)

Mutagenesis of ***Phanerochaete chrysosporium***

A method has been described for the mutagenesis of *P. chrysosporium* using UV or X-ray. The auxotrophic mutants may improve lignocellulose degradation:
Gold, M. H., Chang, T. M., Mayfield, M. B.: Appl. Environ. Microbiol. *44*, 996 (1982)

Biological Delignification of Wood for Paper and Board Making

A patent has described a method to delignify lignocellulosic materials with white-rot fungus *Pleurotus ostreatus* giving a product suitable for paper or board manufacture. The process is less expensive to operate than a chemical pulping process and avoids pollution problems:
Eisenstein, A.: Cellulose Production from Lignocellulosic Materials. Europe Patent No. 60–467, 22. 09. 1982

Biological Detoxification of Kraft Pulp Mill

The biodegradation of the chlorinated resin acid, 14-chlorodehydroabietic acid (14-CDA), found in Kraft pulp mill effluent and known to be toxic to fish, has been studied using fungus *Mortierella isobellina*. The fungus converted the 14-CDA to products of low toxicity to fish:
Kutney, J. P. et al.: Helv. Chem. Acta *65*, 1343 (1982)

Lignin as Binding Material

Polymerization method of lignosulfonates to produce a binder for use in production of chipboard etc. has been patented. White-rot fungus *Fomes annosus* could polymerize lignosulfonate with an average molecular weight of 435000 to a product with a molecular weight of 1500000 in 19 days. This polymer can replace the expensive phenolic resins and isocyanates in chipboards:
Ges. Biotechnol. Forsch.: Phenolic Binder Preparation from Lignin Sulfonate. Germany Patent No. 3037-992. 19. 08. 1982

For the larg-scale application of lignin as wood adhesives and asphalt extenders see:
Sundstrom, D. W., Klei, H. E.: Biotech. Bioeng. Symp. No. 12, 45 (1982)

11 References

1. Abbot, J., Bolker, H. I.: Tappi *65*, 37 (1982)
2. Akin, D. E.: Appl. Environ. Microbiol. *40*, 809 (1980)
3. Alber, W. et al.: Use of lignin on lignin-containing materials, German (DR) Pat. No. 133, 788, 24 Jan. 1979
4. Alcohol Fuels Process R/D Newsletter, US Solar Energy Res. Inst., p 39, Winter 1980
5. Alibert, G., Boudet, A.: Physiol. Vég. *17*, 67 (1979)
6. Allen, B. R., Cousin, M. J., Pierce, G. E.: Pretreatment Methods for the Degradation of Lignin, p. 17, Battelle Columbus Laboratories Report, Ohio 1980
7. Amer, G. I.: Lignin Biodegradation, Reduced Oxygen Species. Ph. D. thesis, Virginia Polytechnic Inst. and State Univ. 1981
8. Amer, G. I.: Drew, S. W.: In: Ann. Rep. Ferment. Processes, Vol. 4 (Tsao, G. T. ed.) p. 67. New York: Academic Press 1980
9. Ander, P., Hatakka, A., Eriksson, K.-E.: Arch. Microbiol. *125*, 189 (1980)
10. Ander, P. et al.: In: The Ekman-Days, Int. Symp. Wood and Pulping Chem., Vol. 3, p. 72, Stockholm: SPCI 1981
11. Antai, S. P., Crawford, D. L.: Appl. Environ. Microbiol. *42*, 378 (1981)
12. Antai, S. P., Crawford, D. L.: Eur. J. Appl. Microbiol. Biotech. *14*, 165 (1982)
13. Azuma, J.-I., Takahashi, N., Koshijima, T.: Carbohydr. Res. *93*, 91 (1981)
14. Bache, R., Pfennig, N.: Arch. Microbiol. *130*, 255 (1981)
15. Bansal, B. B., Hornof, V., Neale, G.: Can. J. Chem. Eng. *57*, 203 (1979)
16. Bar-Lev, S. S., Kirk, T. K.: Biochem. Biophys. Acta, *99*, 373 (1981)
17. Barder, M. J., Crawford, D. L.: Can. J. Microbiol. *27*, 859 (1981)
18. Biomass Digest, *3*, 11, 3 (1981)
19. ibid. *3*, 12, 3 (1981)
20. ibid. *4*, 4, 1 (1982)
21. ibid. *4*, 6, 5 (1982)
22. Björkman, A., Person, B.: Svensk Papperstidn. *60*, 158 (1957)
23. Bollag, J.-M., Liu, S.-Y., Minard, R. D.: Soil Sci. Soc. Am. J. *44*, 52 (1980)
24. Bollag, J.-M., Liu, S.-Y., Minard, R. D.: Soil Biol. Biochem. *14*, 157 (1982)
25. Boutelje, J. B., Eriksson I.: Svensk Papperstidn. *85*, R39 (1982)
26. Brice, R. E., Morrison, I. M.: Carbohydr. Res. *101*, 93 (1982)
27. Brunow, G., Poppius, K.: Acta Chem. Scand. *36*, 377 (1982)
28. Budin, Von D., Susa, L.: Holzforschung *36*, 17 (1982)
29. Buswell, J. A., Eriksson, K.-E., Pettersson, B.: J. Chromatogr. *215*, 99 (1981)
30. Chang, H.-M., Chen, C.-L., Kirk, T. K.: In: Lignin Biodegradation: Microbiology, Chemistry, and Potential Applications (Kirk, T. K., Higuchi, T., Chang, H.-M. eds.) Vol. 1, p. 215. Florida: CRC Press 1980
31. Chem. Eng. News *58*, 44, 35 (1980)
32. Chen, C.-L., Chang, H.-M., Kirk, T. K.: Holzforschung *36*, 3 (1982)
33. Chiwetelu, C., Hornof, V., Neale, G. H.: Trans I. Chem. E. *60*, 177 (1982)
34. Chua, M. G. S., Wayman, M.: can. J. Chem. *57*, 2603 (1979)
35. Chua, M. G. S. et al.: Holzforschung *36*, 165 (1982)
36. Clayton, N. E., Srinivasan, V. R.: Naturwissenschaften *68*, 97 (1981)
37. Colberg, P. J., Young, L. Y.: Can. J. Microbiol. *28*, 886 (1982)
38. Concin, R., Burtscher, E., Bobleter, O.: Holzforschung *35*, 279 (1981)
39. Connors, W. J. Sarkanen, S., Mc Carthy, J. L.: ibid. *34*, 80 (1980)
40. Crawford, D. L.: Biotech. Bioeng. Symp. No. 11, 275 (1981)
41. Crawford, D. L., Crawford, R. L.: Enzyme Microb. Technol. *2*, 11 (1980)
42. Crawford, D. L. et al.: Arch. Microbiol. *131*, 140 (1982)
43. Crawford, R. L. (ed.): Lignin Biodegradation and Transformation. New York: John Wiley 1981
44. Crawford, R. L., Crawford, D. L., Dizikes, G. L.: Arch. Microbiol. *129*, 204 (1981)
45. Crawford, R. L., Robinson, L. E., Foster, R. D.: Appl. Environ. Microbiol. *41*, 1112 (1981)
46. Das, N. N. et al.: Carbohydr. Res. *94*, 73 (1981)

47. Deschamps, A. M., Gillie, J. P., Lebeault, J. M.: Eur. J. Appl. Microbiol. Biotech. *13*, 222 (1981)
48. Deschamps. A. M., Mahoudeau, G., Lebeault, J. M.: ibid. *9*, 45 (1980)
49. Donnelly, M. I., Dagley, S.: J. Bacteriol. *142*, 916 (1980)
50. Draganova, Von R., Nedeltscheva, M., Tzolova, E.: Holzforschung *35*, 223 (1981)
51. Drew, S. W., Kadam, K. L.: Develop. Indust. Microbiol. *20*, 153 (1979)
52. Drew, S. W. et al.: AIChE Symp. Series No. 181, *74*, 21 (1978)
53. Eaton, D., Chang, H.-M., Kirk, T. K.: Tappi *63*, 103 (1980)
54. Eggeling, L., Sahm, H.: Arch. Microbiol. *126*, 141 (1980)
55. Ellwardt, Von P.-Chr., Haider, K., Erst, L.: Holzforschung, *35*, 103 (1981)
56. Enoki, A., Golsby, G. P., Gold, M. H.: Arch. Microbiol. *125*, 227 (1980)
57. Enoki, A., Goldsby, G. P., Gold, M. H.: ibid. *129*, 141 (1981)
58. Enoki, A. et al.: FEMS Microbiol. Lett. *10*, 373 (1981)
59. Eriksson, K.-E., Grünewald, A., Vallander, L.: Biotech. Bioeng. *22*, 363 (1980)
60. Eriksson, K.-E.: In: The Ekman-Days, Int. Symp. Wood and Pulping Chem., Vol. 3, p. 60, Stockholm: SCPI 1981
61. Eriksson, K.-E. et al.: Holzforschung *34*, 207 (1980)
62. Falkehag, S. I.: In: Appl. Polym. Symp. No. 28. p. 247. New York: John Wiley 1975
63. Fenn, P., Choi, S., Kirk, T. K.: Arch. Microbiol. *130*, 66 (1981)
64. Fenn, P., Kirk, T. K.: ibid *123*, 307 (1979)
65. Fenn, P., Kirk, T. K.: ibid. *130*, 59 (1981)
66. Fenner, R. A., Lephardt, J. O.: J. Agric. Food Chem. *29*, 846 (1981)
67. Fischer, F.: Zellstoff und Papier *1*, 13 (1980)
68. Franzidis, J.-P., Porteous, A.: In: Fuels From Biomass and Wastes. (Klass, D. L., Emert, G. H. eds.), p. 267. Kent: Ann Arbor Science Publishers Inc. 1981
69. Freudenberg, K., Friedmann, M.: Chem. Ber. *93*, 2138 (1960)
70. Gadda, L.: Delignification of Wood Fibre Cell Wall during Alkaline Pulping Processes as studied by UV-Microspectrophotometry. Thesis, Inst. of Wood Chemistry and Pulp and Paper Technology of Abo Akademi, Finland 1981
71. Garrod, B. et al.: New Phytol. *90*, 99 (1982)
72. Gellerstedt, G., Agnemo, R.: Acta Chem. Scand. *34*, 461 (1980)
73. Gierer, J.: Wood Sci. Technol. *14*, 241 (1980)
74. Gierer, J.: Holzforschung *36*, 43 (1982)
75. Gierer, J.: ibid. *36*, 55 (1982)
76. Goldsby, G. P., Enoki, A., Gold, M. H.: Arch. Microbiol. *128*, 190 (1980)
77. Grisebach, H.: In: The Biochemistry of Plants (Stumpf, P. K. Conn, E. E. eds.), Vol. 7, p. 457, New York: Academic Press 1981
78. Gross, G. G.: In: Advances in Botanical Res., Vol. 8, p. 25. London: Academic Press 1980
79. Gupta, J. K. et al.: Arch. Microbiol. *128*, 349 (1981)
80. Haars, A., Hüttermann, A.: ibid. *125*, 233 (1980)
81. Haars, A., Hüttermann, A.: Naturwissenschaften *67*, 39 (1980)
82. Haars, A. et al.: Holzforschung *36*, 85 (1982)
83. Hahlbrock, K., Grisebach, H.: Ann. Rev. Plant Physiol. *30*, 105 (1979)
84. Haider, K., Martin, J. P.: Soil. Biol. Biochem. *13*, 447 (1981)
85. Hall, P. L.: Enzyme Microb. Technol. *2*, 170 (1980)
86. Hall, P. L., Glasser, W. G., Drew, S. W.: In: Lignin Biodegradation: Microbiology, Chemistry, and Potential Applications (Kirk, T. K., Higuchi, T., Chang, H.-M. eds.) Vol. 2, p. 33. Florida: CRC Press 1980
87. Haltmeier, T., Leisola, M.: Personal communication, Inst. of Biotechnol. ETH, 8093 Zürich 1982
88. Hanselmann, K. W.: Experientia *38*, 176 (1982)
89. Hatakeyama, T., Hatakeyama, H.: Polymer *23*, 475 (1982)
90. Hayatsu, R. et al.: Nature *278*, 41 (1979)
91. Healy, J. B., Young, L. Y., Reinhard, M.: Appl. Environ. Microbiol. *39*, 436 (1980)
92. Hedges, J. I., Ertel, J. R.: Anal. Chem. *54*, 174 (1982)
93. Hemmingson, J. A., Leary, G.: Aus. J. chem. *33*, 917 (1980)
94. Highley, T. L.: Can. J. For. Res. *12*, 435 (1982)

95. Higuchi, T.: Wood Res. *66*, 1 (1980)
96. Higuchi, T.: In: The Ekman-Days, Int. Symp. Wood and Pulping Chem., Vol. 3, p. 16, Stockholm: SPCI 1981
97. Higuchi, T.: Wood Res. *67*, 47 (1981)
98. Higuchi, T.: Experimenta *38*, 159 (1982)
99. Higuchi, T., Nakatsubo, F.: Kemia-Kemi *9*, 481 (1980)
100. Hüttermann, A., Herche, C., Haars, A.: Holzforschung *34*, 64 (1980)
101. IEA, Biomass Conversion Technol. Inform. Service p. 125 No. 413, Dublin, June 1980
102. Iwahara, S., Ishiki, K., Higuchi, T.: Nippon Nôgeikagaku Kaishi *55*, 991 (1981)
103. Iwahara, S., Koaka, H.: Technical Bull. of Faculty of Agriculture, Kagawa Univ. *33*, 7 (1981)
104. Iwahara, S. et al.: J. Ferm. Technol. *58*, 183 (1980)
105. Janshekar, H., Brown, C., Fiechter, A.: Anal. Chim. Acta *130*, 81 (1981)
106. Janshekar, H., Fiechter, A.: Eur. J. Appl. Microbiol. Biotech. *14*, 47 (1982)
107. Janshekar, H., Haltmeier, T., Brown, C.: ibid. *14*, 174 (1982)
108. Janshekar, H. et al.: Arch. Microbiol. *132*, 14 (1982)
109. Jeffries, T. W., Choi, S., Kirk, T. K.: Appl. Environ. Microbiol. *42*, 290 (1981)
110. Joseleau, J.-P., Gancet, C.: Svensk Papperstidn. *84*, R 123 (1981)
111. Kaiser, J.-P., Hanselmann, K. W.: Experientia *38*, 167 (1982)
112. Kamaya, Y. et al.: Arch Microbiol. *129*, 305 (1981)
113. Kaplan, D.: Phytochem. *18*, 1917 (1979)
114. Katayama, T., Higuchi, T.: 25th Symp. Lignin Chemistry, Fukuoka, Japan 1980
115. Katayama, T., Nakatsubo, F., Higuchi, T.: Arch. Microbiol. *126*, 127 (1980)
116. Katayama, T., Nakatsubo, F., Higuchi, T.: ibid. *130*, 198 (1981)
117. Kern, H. W.: In: Microbiol Degradation of Xenobiotics and Recalcitrant Compounds. (Leisinger, T. et al. eds.) p. 299. New York: Academic Press 1981
118. Keyser, P., Kirk, T. K., Zeikus, J. G.: J. Bacteriol. *135*, 790 (1978)
119. Kirk, T. K. In: Trends in the Biology of Fermentations for Fuels and chemicals (Hollaender, A. ed.), New York: Plenum 1981
120. Kirk, T. K., Fenn, P.: In: Decomposition by basidiomycetes (Hedgar, J., Frankland, J. eds.), p. 67. Cambridge: Cambridge Univ. Press 1981
121. Kirk, T. K., Higuchi, T., Chang, H.-M. (eds.): Lignin Biodegradation: Microbiology, Chemistry, and Potential Applications. Vols. 1 + 2, Florida: CRC Press 1980
122. Kolar, J. J., Lindgren, B. O., Treiber, E.: Svensk Papperstidn. *85*, R21 (1982)
123. Koukios, E. G., Valkanas, G. N.: Ind. Eng. Chem. Prod. Res. Dev. *21*, 309 (1982)
124. Kringstad, K.: The challenge of lignin, In: Proc. World Conf. on Future Sources of Organic Raw Materials (Pierre, L. E. St., Brown, G. R., eds.) p. 627. Oxford: Pergamon Press 1980
125. Kristersson, P., Lundquist, K.: Acta Chem. Scand. *34*, 213 (1980)
126. Kuč, J., Hammerschmidt, R.: Physiol. Plant Pathol. *20*, 61 (1982)
127. Kuhlman, E. G.: Can. J. Bot. *58*, 36 (1980)
128. Kutsuki, H., Higuchi, T.: Planta *152*, 365 (1981)
129. Kutsuki, H., Nakatsubo, F., Higuchi, T.: Mokuzai Gakkaishi *27*, 520 (1981)
130. Kutsuki, H., Shimada, M., Higuchi, T.: Phytochemistry *21*, 19 (1982)
131. Kutsuki, H., Shimada, M., Higuchi, T.: ibid. *21*, 267 (1982)
132. Kuwahara, M., Endo, Y., Noborihayashi, K.: J. Ferment. Technol. *59*, 15 (1981)
133. Lawrence, J.: Water Res. *14*, 373 (1980)
134. Lawrence, L. L., Sally, A. G., Kirk, T. K.: Holzforschung *35*, 67 (1981)
135. Leary, G. J.: Wood Sci. Technol. *14*, 21 (1980)
136. Leary, G. J.: ibid. *16*, 67 (1982)
137. Lindström, T.: Separation Sci. Technol. *14*, 601 (1979)
138. Liu, S.-Y., Minard, R. D., Bollag, J.-M.: Soil Sci. Soc. Am. J. *45*, 1100 (1981)
139. Lora, J. H., Wayman, M.: J. Appl. Polym. Sci. *25*, 589 (1980)
140. Lora, J. H., Wayman, M.: Can. J. Chem. *58*, 669 (1980)
141. Loubinoux, B. et al.: Tetrahedron Lett. *21*, 4991 (1980)
142. Lüderitz, T., Grisebach, H.: Eur. J. Biochem. *119*, 115 (1981)
143. Lüderitz, T., Schatz G., Grisebach, H.: ibid. *123*, 583 (1982)
144. Lundquist, K.: Acta. Chem. Scand. *33*, 27 (1979)

145. Lundquist, K.: ibid. *34*, 21 (1980)
146. Lundquist, K.: ibid. *35*, 497 (1981)
147. Lundquist, K., Josefsson, B., Nyquist, G.: Holzforschung *32*, 27 (1978)
148. Lundquist, K., Kirk, T. K.: Tappi *63*, 80 (1980)
149. Lundquist, K., Simonson, R., Tingsvik, K.: Svensk Papperstidn. *82*, 272 (1979)
150. Lundquist, K., Simonson, R., Tingsvik, K.: Paperi ja Puu, *11*, 709 (1981)
151. Maccubbin, A. E., Hodson, R. E.: Appl. Environ. Microbiol. *40*, 735 (1980)
152. Mackay, D. M., Roberts, P. V.: Carbon *20*, 87 (1982)
153. Martin et al.: Soil Biol. Biochem. *14*, 289 (1982)
154. Matsuda, S., Hiraki, K., Nishikawa, Y.: Bunseki Kagaku *28*, 341 (1979)
155. Maule, A. J., Ride, J. P.: Physiological Plant Path. *20*, 235 (1982)
156. Milstein, O. et al.: Eur. J. Appl. Microbiol. Biotech. *13*, 117 (1981)
157. Monties, B. et al.: Holzforschung *35*, 217 (1981)
158. Muindi, P. J., Thomke, S.: Anim. Feed Sci. Technol. *6*, 197 (1981)
159. Muntifering, R. B., de Gregorio, R. M., Deetz, L. E.: Nutrition Reports International *24*, 543 (1981)
160. Nakatsubo, F.: Wood Res. *67*, 59 (1981)
161. Nakatsubo, F., Reid, I. D., Kirk, T. K.: Biochim. Biophys. Res. Comm. *102*, 484 (1981)
162. Nakatsubo, F. et al.: Arch. Microbiol. *128*, 416 (1981)
163. Namba, H., Nakatsubo, F., Higuchi, T.: 25th Symp. Lignin Chemistry, Fukuoka, Japan 1980
164. Neale, G., Hornof, V., Chiwetelu, C.: Can. J. Chem. *59*, 1938 (1981)
165. Neilson, A. et al.: IVL, Swedish water and Pollution Res. Inst., Report No. B-660, Stockholm 1981
166. Nicholson, J. W. G.: Agric. Environm. *6*, 205 (1981)
167. Nimz, H. H.: Wood Sci. Technol. *15*, 311 (1981)
168. Noguchi, A., Shimada, M., Higuchi, T.: Holzforschung *34*, 86 (1980)
169. Norris, D. M.: Appl. Environ. Microbiol. *40*, 376 (1980)
170. Obst, J.-R.: Holzforschung *36*, 143 (1982)
171. Obst, J.-R.: Tappi *65*, 109 (1982)
172. Obst, J.-R., Sanyer, N.: ibid. *63*, 111 (1980)
173. Odier, E.: In: Proc. Colloque Cellulol. Microb. p. 101, Marseille 1980
174. Odier, E., Janin, G., Monties, B.: Appl. Environ. Microbiol. *41*, 337 (1981)
175. Ohta, M., Higuchi, T., Iwahara, S.: Arch. Microbiol. *121*, 23 (1979)
176. Papadopoulos, J., Chen, C.-L., Goldstein, I. S.: Holzforschung *35*, 283 (1981)
177. Pearce, R. B., Ride, J. P.: Physiol. Plant Pathol. *20*, 119 (1982)
178. Perineau, F., Gaset, A.: J. Chem. Tech. Biotech. *31*, 711 (1981)
179. Philippou, J. L., Johns, W. E., Nguyen, T.: Holzforschung *36*, 37 (1982)
180. Philippou, J. L. et al.: Forest Prod. J. *32*, 27 (1982)
181. Philippou, J. L. et al.: ibid. *32*, 55 (1982)
182. Platt, M. W., Chet, I., Henis, Y.: Eur. J. Appl. Microbiol. Biotech. *13*, 194 (1981)
183. Polcin, J.: In: Proc. IUPAC 27th Int. Congr. Pure and Applied Chemistry, (Varmavuozi, A. ed.). Oxford: Pergamon Press 1980
184. Pometto III, A. L., Crawford, D. L.: Enzyme Microb. Technol. *3*, 73 (1981)
185. Pometto III, A. L., Sutherland, J. B., Crawford, D. L.: Can. J. Microbiol. *27*, 636 (1981)
186. Rast, H. G. et al.: FEMS Microbiol. Lett. *8*, 259 (1980)
187. Reid, I. D.: Can. J. Bot. *57*, 2050 (1979)
188. Reid, I. D., Seifert, K. A.: Can. J. Microbiol. *26*, 1168 (1980)
189. Reid, I. D., Seifert, K. A.: Can. J. Bot. *60*, 252 (1982)
190. Rosenberg, S. L.: Mycologia *72*, 798 (1980)
191. Rotstein, O. D. et al.: Gastroenterology *81*, 1098 (1981)
192. Ruel, K., Barnoud, F., Eriksson, K.-E.: Holzforschung *35*, 157 (1981)
193. Saka, S., Thomas, R. J.: Wood Sci. Technol. *16*, 1 (1982)
194. Saka, S., Thomas, R. J.: Wood and Fiber *14*, 144 (1982)
195. Saka, S. et al.: Wood Sci. Technol. *16*, 139 (1982)
196. Sakakibara, A.: ibid. *14*, 89 (1980)

197. Salkinoja-Salonen, M., Sundman, V.: In: Lignin Biodegradation: Microbiology, Chemistry, and Potential Applications (Kirk, T. K., Higuchi, T., Chang, H.-M. eds.), Vol. 2, p. 179. Florida: CRC Press 1980
198. Samuelson, O., Sjöberg, L.-A.: Svensk Papperstind. *85*, R69 (1982)
199. Sanford, M. E., Detroit, W. J.: Paper presented at the 184th Nat. ACS Meet., Kansas City, Sept. 13, 1982
200. Sarkanen, K. V.: In: Proc. IUPAC 27th Int. Congr. Pure and Applied Chemistry, (Varmavuori, A. ed.), p. 299. Oxford: Pergamon Press 1980
201. Sarkanen, K. V.: In: Proc. OECD/COST Workshop on Improved utilization of Lignocellulosic Materials for Animal Feed (Domsch, K. H., Ferranti, M. P., Theander, O. eds.), p. 19. Braunschweig 1981
202. Sarkanen, S. et al.: Macromolecules *14*, 426 (1981)
203. Schäfer, J. et al.: ibid. *14*, 557 (1981)
204. Schmid, G., Grisebach, H.: Eur. J. Biochem. *123*, 363 (1982)
205. Schmidt, O., Bauch, J.: Wood Sci. Technol. *14*, 229 (1980)
206. Schultz, T. P. et al.: J. Chrom. Sci. *19*, 235 (1981)
207. Schultz, T. P. et al.: ibid. *19*, 235 (1981)
208. Schulz, K. R., Lentz, H., Ziechmann, W.: Erdöl und Kohle — Erdgas — Petrochemie vereinigt mit Brennstoff-Chemie *33*, 42 (1980)
209. Selke, S. M. et al.: Ind. Eng. Chem. Prod. Res. Dev. *21*, 11 (1982)
210. Shimada, M. et al.: Arch. Microbiol. *129*, 321 (1981)
211. Steinitz, Y. L.: Eur. J. Appl. Microbiol. Biotech. *13*, 216 (1981)
212. Sudo, K., Mullord, D. J., Pepper, J. M.: Can. J. Chem. *59*, 1028 (1981)
213. Sudo, K., Pepper, J. M.: ibid. *60*, 229 (1982)
214. Suflita, J. M., Bollag, J.-M.: Soil Sci. Soc. Am. J. *45*, 297 (1981)
215. Sundman, G., Kirk, T. K., Chang, H.-M.: Tappi *64*, 145 (1981)
216. Sundstrom, D. W., Klei, H. E.: Paper presented at the 4th Symp. on Biotech. in Energy Production and Conservation, May 11-14, Tennessee, U.S.A. Abstracts, p. 6, 1982
217. Sutherland, J. B., Crawford, D. L., Speedie, M. K.: Mycologia *74*, 511 (1982)
218. Thiago, L. R. L. de S., Kellaway, R. C.: Anim. Feed Sci. Technol. *7*, 71 (1982)
219. Trenck, K. T. v. d., Hunkler, D., Sandermann, Jr., H.: Z. Naturforsch. *36* C, 714 (1981)
220. Umezawa, T., Nakatsubo, F., Higuchi, T.: Arch. Microbiol. *131*, 124 (1982)
221. Vohra, R. M. et al.: Biotech. Bioeng. *22*, 1497 (1980)
222. Walden, C. C.: Adv. Biotechnol. *2*, 669 (1981)
223. Wayman, M., Chua, M. G. S.: Can. J. Chem. *57*, 2599 (1979)
224. Wegener, G.: Holz als Roh- und Werkstoff *40*, 209 (1982)
225. Weichelt, T.: Z. Pflanzenernähr. Bodenk. *144*, 565 (1981)
226. Weichelt, T.: Z. Acker- und Pflanzenbau *150*, 480 (1981)
227. Weichelt, T.: z. Pflanzenernähr. Bodenk. *145*, 42 (1982)
228. Weinstein, D. A. et al.: Appl. Environ. Microbiol. *39*, 535 (1980)
229. Werthemann, D. P.: Tappi *65*, 98 (1982)
230. Whiting, P., Goring, D. A. I.: Svensk Papperstidn. *84*, R120 (1981)
231. Whitmore, F.: Phytochemistry *21*, 315 (1982)
232. Wicklow, D. T., Detroy, R. W., Adams,, S.: Mycologia *72*, 1065 (1980)
233. Wu. L. C.-F., Glasser, W. G.: Biotech. Bioeng. *21*, 1679 (1979)
234. Yan, J. F.: Science *215*, 1390 (1982)
235. Yan, J. F., Johnson, D. C.: J. Agric. Food Chem. *28*, 850 (1980)
236. Yang, H. H., Effland, M. J., Kirk, T. K.: Biotech. Bioeng. *22*, 65 (1980)
237. Yang, J.-M., Goring, D. A. I.: Can. J. Chem. *58*, 2411 (1980)
238. Zadrazil, F., Brunnert, H.: Eur. J. Appl. Microbiol. Biotech. *9*, 37 (1980)
239. Zadrazil, F., Brunnert, H.: Eup. J. Appl. Microbiol. Biotech. *11*, 183 (1981)
240. Zeikus, J. G.: In: Lignin Biodegradation: Microbiology, Chemistry, and Potential Applications (Kirk, T. K., Higuchi, T., Chang, H.-M. eds.), Vol. 1, p. 101. Florida: CRC Press 1980
241. Zeikus, J. G.: In: Adv. Microbiol. Ecology (Alexander, M. ed.), *5*, p. 211. New York: Plenum Publ. Corp. 1981
242. Zemek, J. et al.: Folia Microbiol. *24*, 483 (1979)

www.ingramcontent.com/pod-product-compliance
Lightning Source LLC
LaVergne TN
LVHW060622300926
849162LV00008B/1688